Histories of Human Engineering

The dream of control over human behaviour is an old dream, shared by many cultures. This fascinating account of the histories of human engineering describes how technologies of managing individuals and groups were developed from the nineteenth century to the present day, ranging from brainwashing and mind control to Dale Carnegie's art of dealing with people. Derksen reveals that common to all of them is the perpetual tension between the desire to control people's behaviour and the resistance this provokes. Thus, to influence other people successfully, technology had to be combined with tact: with a personal touch, with a subtle hint or with outright deception, manipulations were made palatable or invisible. Combining psychological history and theory with insights from science and technology studies and rhetorical scholarship, Derksen offers a fresh perspective on human engineering that will appeal to those interested in the history of psychology and the history of technology.

Maarten Derksen is Assistant Professor in the Faculty of Behavioural and Social Sciences at the University of Groningen. He has authored *Iedereen doet aan psychologie* (1999) and, with Sybe Terwee, translated Wittgenstein's *Philosophical Investigations* into Dutch (1992; 2nd ed. 2006). He was an honorary fellow of the Science Studies Centre of the University of Bath, an associate of the 'What Makes Organization?' research programme at the Copenhagen Business School, and is currently a member of the editorial board of *Theory & Psychology*.

Histories of Human Engineering

Tact and Technology

Maarten Derksen
University of Groningen

CAMBRIDGE
UNIVERSITY PRESS

CAMBRIDGE
UNIVERSITY PRESS

University Printing House, Cambridge CB2 8BS, United Kingdom

One Liberty Plaza, 20th Floor, New York, NY 10006, USA

477 Williamstown Road, Port Melbourne, VIC 3207, Australia

4843/24, 2nd Floor, Ansari Road, Daryaganj, Delhi – 110002, India

79 Anson Road, #06–04/06, Singapore 079906

Cambridge University Press is part of the University of Cambridge.

It furthers the University's mission by disseminating knowledge in the pursuit of education, learning, and research at the highest international levels of excellence.

www.cambridge.org
Information on this title: www.cambridge.org/9781107057432
DOI: 10.1017/9781107414921

First published 2017

Printed in the United Kingdom by Clays, St Ives plc

A catalogue record for this publication is available from the British Library.

ISBN 978-1-107-05743-2 Hardback

Contents

Acknowledgements

Chapter 3 is a revised version of 'Turning Men into Machines? Scientific Management, Industrial Psychology, and the "Human Factor"' (Derksen, Maarten; *Journal of the History of the Behavioral Sciences* 50, no. 2 (2014): 148–65). I thank Signe Vikkelsø for her detailed and constructive review; reviewer 2 for at least taking the time to read the manuscript and saying a few words; the editor, Ian Nicholson, for guiding the manuscript to publication; and the publisher, John Wiley and Sons, for giving permission to use the article here.

Chapter 4 is a revision of 'The History of "Social Technology", 1898–1930' (Derksen, Maarten and Wierenga, Tjardie; *History and Technology* 29, no. 4 (2013): 311–30), which can be found at www.tandfonline.com. I thank the editor, Martin Collins, for his expert handling of the manuscript; the two reviewers for their helpful reviews; the publisher, Taylor & Francis, for giving permission to use the article here; and above all Tjardie Wierenga for working on it with me and allowing me to use it for this book.

The project of which this book is a product has its origin in a dinner conversation years ago, when Anne Beaulieu and I discovered we were both interested in social technology. As so often in our life together, Anne turned what would otherwise have remained merely an interesting idea into a project we could work on together. My greatest debt is to her.

Without Hetty Marx, formerly commissioning editor at Cambridge University Press, there would have been no book. I'm still deeply embarrassed that she had to remind me twice that I had promised to write a proposal. I hope the final product lives up to the faith she had in me. I thank the reviewers for their comments on the proposal and the first draft. When Hetty left to become an independent editor Janka Romero took over, and she has been a great help.

I'm very grateful to Liz Barton Royer and the staff of the Drs. Nicholas and Dorothy Cummings Center for the History of Psychology at the University of Akron, Ohio, for their assistance during my visit there. At my own faculty library Sander Sprik was invaluable.

Over the years, I have benefitted from conversations with many colleagues. Anne Beaulieu, Signe Vikkelsø and I organised a workshop in 2009 on social technology, generously funded by WTMC, the Netherlands Graduate Research School of Science, Technology and Modern Culture. Our discussions with Katja Mayer, Tereza Stöckelová, Kathrin Braun, Javier Lezaun, Ingmar Lippert, Jonna Brenninkmeijer, Katia Dupret Søndergaard, Julie Sommerlund, Sara Malou Strandvad, Paul Wouters and Bas van Heur were a great joy and inspiration. Katja pointed me to Charles Henderson and his social technology and thus made me realise that the concept has a history. Steve Brown couldn't make it, but his work is always important to me.

Paul du Gay and Signe Vikkelsø were kind enough to have me over for a few days in 2011 at the Department of Organization of the Copenhagen Business School, where I had inspiring discussions about organisation and management with them and with Karen Boll, Chris Grey and Chris Mathieu and, at the Department of Management, Politics and Philosophy, with Kaspar Villadsen.

I have the good fortune to work in the world's best Theory and History of Psychology group. Over the years my ideas and drafts were listened to, read and commented on by Jonna Brenninkmeijer, Adeena Mey, Hilde Tjeerdema, Jess Cadwallader, Felix Schirmann, Trudy Dehue, Carlos Baum, Stephan Schleim, Berend Verhoeff and Douwe Draaisma. One way or another, they were all important to me, and I love them all, but my greatest debt is to Douwe, who read my original proposal and all the chapters and offered invaluable suggestions. If this book comes anywhere near the clarity and insight of his work, readers may count themselves lucky.

Dennis Bryson prompted me to clarify why I think the difference between 'social technology' and 'social engineering' is significant. My ideas about replication were much influenced by my collaboration with Eric Rietzschel and by an extensive email conversation with Hans Radder. Nathaniel Rivers taught me a lot about rhetoric, *kairos* in particular. My friend Steve Course, somewhat worryingly, turned out to be an expert on the CIA; he read the chapter about brainwashing and offered many valuable suggestions. (He is convinced Dulles was in on the secret, by the way.) Marcia Holmes read the chapter too and sent me extensive, helpful comments.

I am immensely grateful to all these people. No doubt many of them will be surprised by what, if anything, came of their advice. Perhaps others will, upon finishing this book, wish I had consulted them too, thinking that might have done some good. In either case, I accept full responsibility for all of the book's flaws.

1 Introduction

A Revolution in Marketing

A revolution has been proclaimed in the field of marketing: the age of neuromarketing is dawning. Now that researchers can look directly into the brain with an fMRI scanner, they can see what really attracts people in an advertisement or television commercial and what moves them to buy a product. Neuromarketing research has found our 'buy button', and if you have something to sell, it can tell you exactly how to push that button. How it works is described in a book aimed at business owners, salespeople and marketers titled *The Buy Button.*[1] It was written by Martin de Munnik, one of the people behind Neurensics, an Amsterdam research and consulting firm that specialises in neuromarketing. In *The Buy Button*, subtitled *The Secret of the Consumer's Brain*, De Munnik tells us of the groundbreaking discovery that the decision to buy a product is the result of the interplay between three brain areas: the nucleus accumbens, the insula and the medial prefrontal cortex. These three areas together are the buy button, 'the greatest neuromarketing revelation of all time', as the book's cover proclaims. Companies like Neurensics have taken this scientific discovery and turned it into practical applications. With an fMRI scanner they can look into the brain and see the activity in these areas when subjects are presented with, for example, different advertisements for breakfast cereal or nail polish. Choosing the best advertisement, once an art based on gut feelings, has become a science. Marketing has been transformed into a 'lottery without blanks'.[2] That, at least, is the claim.

Let us for the moment put aside the question of whether there really is a buy button in the brain – suffice it to say that there are other neuromarketing researchers who call the idea 'bullshit' and warn of 'neurocowboys' harming the field by overselling it.[3] I want to focus on the dream that the buy button stands for: the dream of knowing what really moves people and using that knowledge to manipulate their behaviour in an effective and efficient manner. The metaphor of a button in the brain that triggers an urge to buy a product represents an ideal of control over human behaviour on a par with the control we have over material devices. Push a button and the consumer wants a cup of coffee, push another button and the espresso machine produces a ristretto. If you can find the buttons, human behaviour is as predictable and as controllable as an espresso machine, albeit more complicated. Given the right knowledge, human behaviour can be engineered.

The dream of control over human behaviour is an old dream, found in many cultures. The buy button has a long history behind it. Control over people has been pursued with magical spells and love potions, hypnosis and charisma, education and propaganda. Nowadays, promises of control often involve the brain. In fact, the story goes, only now that we have shifted our attention from the mind to the brain and are beginning to understand how the brain works are we starting to realise the dream.[4] That is the revolution: after centuries of confusion, we at last begin to understand the mechanism behind human behaviour because we're finally able to see what happens in the brain. Victor Lamme, professor of cognitive neuroscience at the University of Amsterdam and founding director of Neurensics, put the idea in the clearest terms in his book *Free Will Doesn't Exist.*[5] Modern neuroscience has revealed that free will is an illusion. The rational self that thinks it is in control, making decisions after internal deliberation, looking before it leaps, is a mere chatterbox that produces a mental monologue as it desperately tries to rationalise the decisions taken unconsciously by other processes in the brain. To influence people's behaviour, we mustn't talk to them (their conscious selves), we must influence their brains. The relevance of this fact goes beyond the field of marketing. Our justice system too should shed its belief in free will and the focus on retribution that comes with it. No crime is pre-meditated because none of our behaviour is pre-meditated. The judgements of a court of law should be based on strictly utilitarian principles and objective knowledge about the brain. Their only aim should be to make sure that this individual, with this brain, does not commit more crimes. Correction, rather than punishment, should be the goal.

For De Munnik and Lamme, and many others worldwide who either work in neuroscience or are impressed by its progress and promise, the ideal of engineering human behaviour revolves around the brain. The same ideal is also expressed in another way, however, that does not involve the brain, at least not directly. Ap Dijksterhuis is a professor of social psychology at the Radboud University in Nijmegen, and he is also one of the founding directors of D&B, a research and consulting firm specialising in behaviour modification. Dijksterhuis rose to international fame in the late 1990s with an experimental study of a so-called 'priming' effect. Research subjects who had been asked to imagine a typical professor and list the 'behaviors, lifestyle, and appearance attributes' of such a professor later scored higher on a general knowledge test than subjects who had imagined and described a secretary.[6] Activating (priming) the mental representation of a typical professor produced behaviour that corresponded with that stereotype. There was no conscious effort involved, however. None of the subjects had been aware of a relation between the prime and the test, and many expressed surprise that the experimenter thought the prime might have raised their scores. Priming works unconsciously, automatically.

Researchers like Dijksterhuis and John Bargh, the pioneer of this kind of priming research, claim that such unconscious influences far outweigh what conscious control we have over our behaviour. What we do is not determined by our conscious thoughts but by unconscious forces. Dijksterhuis's company D&B applies this knowledge to develop 'interventions' that influence the behaviour of 'citizens, consumers, or employees'.[7] Like Lamme and Neurensics, D&B bypasses consciousness and directly targets the mechanisms of behaviour. Because most of our behaviour is automatic and directed by unconscious impulses, asking people questions in surveys and interviews is of very little use in finding out what drives them. If you want to know what consumers think of your brand or product, you must measure their unconscious associations and actual behaviour. If you want to change people's behaviour you need to do more than 'make them conscious' of its dangers. Effective behavioural change requires more subtle techniques.

In the way they present themselves, their ideas and methods, their achievements and promises, Neurensics and D&B project a similar ideal of human engineering, the outlines of which are drawn by the contrast between the illusions of consciousness and the reality of unconscious mechanisms, between ineffective talk and efficient, proven techniques. Firstly, both companies offer what sociologists of science Tanja Schneider and Steve Woolgar have called 'technologies of ironic revelation', ways of revealing a truth that is hidden behind our everyday illusions about ourselves.[8] Consumers are pictured as fundamentally and inevitably mistaken about themselves. They do not know and cannot know what moves them, although they think they do.[9] This is an old motif in the history of psychology, a discipline that has often felt the need to assert its cognitive authority by arguing lay people's essential ignorance and incompetence in psychological matters.[10] Nothing shows the superiority of psychology and neuroscience better than experimental evidence that not only are ordinary people irrational, but they're not even aware that they are, and, to top it off, they couldn't be aware of it because the processes that determine their behaviour take place unconsciously, in their brain. Science is superior on three levels.

Secondly, what Neurensics and D&B offer is not only a technology of revelation but also one of production: knowing the true mechanism behind consumer behaviour allows modifying that behaviour efficiently and effectively. We could call it a technology of ironic manipulation because it is precisely because people are unaware of what moves them that the neuromarketer and behavioural scientist can manipulate their behaviour so well. The very fact that the real springs of action lie in the mechanisms of the brain rather than in conscious deliberation and free will makes this behavioural engineering so powerful.

Realising their compatibility, Neurensics and D&B have become partners. In a press release on July 6, 2012, 'the two technology leaders in neuroeconomics' announced their alliance. The fruit of their partnership, a unique combination of imaging techniques and behavioural change that unites the biology and psychology of behaviour, will be 'a management tool of unparalleled power'.[11]

Resistance

This book is about the dream of effective and efficient control over human behaviour and society, about attempts to make it reality, but also about the resistance it encounters. Not everyone likes the idea of controlling people. What to some is a dream to others is a nightmare. They consider it wrong, immoral, inhuman to seek control over others. To be fully human is to be autonomous; by manipulating people you degrade them to the level of slave, animal or machine. Immanuel Kant, for example, strongly disapproved of rhetoric, which he considered to be 'an insidious art that knows how, in matters of moment, to move men like machines'.[12] Kant ascribed to rhetoric the power to deprive people of what he thought was the essence of human beings, their reason and autonomy. It is an art that merits no respect.

Control over people encounters resistance not only as an idea but also in practice. People tend to resist attempts to control and manipulate them. They usually object to being treated as objects. To be effective and efficient, therefore, techniques of control have to take that resistance into account. Ways of preventing or circumventing people's recalcitrance must be designed into them. Neurensics claims to achieve this by directly targeting the brain, avoiding whatever little conscious control people have over their behaviour. It is this relation between techniques of control and the resistance to them that I focus on in this book. As an example of that interplay, consider again the work of D&B.

Despite the partnership and the shared ideal of power, efficiency and a way beyond talk, the claims of D&B are noticeably more cautious and nuanced than those of Neurensics. There is, for instance, no equivalent of the 'buy button' in their description of the mechanisms of human behaviour, and their rejection of consciousness is less than total. Although influencing behaviour requires more than changing people's minds, techniques of conscious influence can be part of effective methods of behavioural change. All in all, human engineering is a more complicated affair for D&B than it is for Neurensics. Many of the complications revolve around the problem of resistance.

One reason that subtle techniques are required to change people's behaviour is that people don't like to change their ways, particularly if someone else tells them to. Overt attempts to convince them of the need to act differently only raise resistance. As Rick van Baaren (the B of D&B) explains in an interview with the in-house magazine of the Dutch Tax Office, 'people have a natural resistance to change'.[13] Acknowledging the unease among employees about impending organisational changes is an important first step towards neutralising it before it festers and erupts into conflict. Next, more subtle techniques may be required, such as 'altercasting': employees who are enthusiastic about the changes are given a role in the process with a fancy label such as 'innovator'. Their confidence thus raised, they will become ambassadors for the plans, spreading their own enthusiasm among their colleagues. Van Baaren immediately adds a note of caution: such techniques should never be employed as simple tricks, that is, without a proper analysis of the situation and the resistance among the employees. 'I want to talk about that resistance, I want to know *why* someone feels discontent.'[14]

The possibility of resistance complicates behavioural change beyond the straightforward pushing of buttons that Neurensics offers its clients. Resistance is, first of all, the raison d'être of this technology. It is not only people's natural inertia as creatures of habit that requires skill to overcome but also the fact that people do not like being told what to do. Subtle techniques of influence are needed because change is not just hard, it is often actively resisted. In the world of D&B, human beings are at once complicated mechanisms as well as stubborn individuals with a will of their own. Consciousness is presented on the one hand as a theatre of illusions, with the illusion of control as the central mistake, but on the other hand the target's consciousness can be a nuisance and it is better to let sleeping dogs lie when attempting to influence people's behaviour. Secondly, resistance is also countered directly by techniques that aim to neutralise it. There is a set of techniques in the behavioural change repertoire specially designed not to produce actual change but to clear the way for it. One such technique is 'acknowledging resistance', mentioned above by Van Baaren, and here we may note the paradox that resistance can apparently be combated by drawing attention to it.

In Pacelle van Goethem's *Selling Ice to Eskimos*, a book that her own website calls 'the Dutch reference work about persuasion',[15] the relation between control and resistance is explained in terms of the brain. Van Goethem is a successful coach, consultant and author on persuasion and influence who has 'discovered the laws of persuasion'.[16] The laws are rooted in the brain. Three brain areas play a role in persuasion: the pre-frontal cortex, the amygdala and the hippocampus. The pre-frontal cortex is the seat of thought and planning but also of self-control. The amygdala is our 'centre for motivation and stress'.[17] It is particularly attuned to negative events and prone to panic (fight, flight or freeze) when we encounter them. Fortunately, the cortex has a direct link with the amygdala. Particularly important for self-control is the ratio of the orbito-frontal cortex to the amygdala: the bigger the former is in relation to the latter, the more you are able to control your impulses. Finally, the hippocampus is involved in memory. When it recognises a situation, particularly when the situation triggers pleasant memories, the hippocampus activates the pituitary gland to release hormones that make us relax and feel good, like oxytocin, dopamine and endorphins. These hormones in turn make the prefrontal cortex work harder and better.

The interplay between these three structures gives us the key to persuasion: relaxation. In situations that are or seem familiar, with people we trust and who appear to give us what we want, we relax and become docile, unless the issue at hand is one we are interested in. In that case, our pre-frontal cortex is alerted and we start to think, aided by the hormones from our pituitary gland. Then we can only be persuaded by good arguments, lest we become recalcitrant. What must be avoided at all costs is to trigger the amygdala into a stress response. When that happens, we are full of resistance (fight, flight or freeze) and persuasion is impossible.

Van Goethem summarises it all in a 'Model of Influence©', a diagram with two dimensions: relaxation versus resistance and low versus high commitment.[18] There are thus four kinds of influence. When we are relaxed but not committed to the issue, we tend to meekly and unconsciously do what we are told – we 'follow'. If we are both relaxed and committed, we react with conscious thought. If we are not relaxed but not interested either, we react with passive resistance – we 'drop out'. Finally, if we are not relaxed but are highly committed to the issue, we react with active resistance – we 'fight'. Persuasion, according to Van Goethem, is thus first of all a matter of inducing relaxation and avoiding resistance. To relax our audience, we must appease their brains (their amygdalas, hippocampuses and prefrontal cortices) by giving them what they want and expect. The best way to do that is with personal persuasiveness, which comes in three styles: the friend, the authority and the example. Each style inspires trust in a different way, but they all have their drawbacks as well, and if your style is not appropriate to the situation and the audience, if what it gives does not match what the brains want and expect, it will meet resistance. If, for example, the situation calls for an 'authority', the 'friend' is a nuisance.

Fortunately, there is a solution to this conundrum that 'really works': to combine the three styles.[19] 'The more of all three styles you radiate towards the other(s), the more persuasive you are.'[20] If people see you as authority, friend *and* example, you are 'maximally persuasive, perhaps even irresistible'.[21] Barack Obama, for example, is almost universally considered as an authority, friend and example. He has maximised his persuasiveness.[22] However, although combining the three styles allows the persuader to cover all bases and be maximally persuasive in all situations, for any audience, judgement remains required. For example, high-status behaviour is the primary element of the 'authority' style. It is such a powerful technique that it trumps all the others – it alone is enough to 'reach the top'.[23] High-status behaviour consists of 'acting as if you feel at home' and 'taking up space', but it needs to be appropriate to the situation: 'The more people are listening to you, the more space you need to fill.'[24]

Like Van Baaren, Van Goethem combines a conception of persuasion as a potentially immensely powerful, effective way of getting people to do what you want with an emphasis on the situatedness of persuasion. They both present a technology based on universal, scientific knowledge (the laws of persuasion, the functioning of the brain) that gives great power to influence people's behaviour and opinions. However, to be effective these technologies also require an understanding of the particular situation in which they are deployed: of the wants and needs of the audience that you're addressing, of the atmosphere in the company that is changing its organisation. Resistance is a potentially disruptive element of every situation, not only because people are generally difficult to get into motion and change their ways (in that sense the point of any behavioural technology is to overcome inertia) but also because people are prone to resist if they're not handled well. The methods that are used to effect change in behaviour or opinions need to be adapted to the people that they target and the situation they are in, lest they are resisted. In short, these methods need to be applied with tact.

Tact and Technology

In *The Practice of Everyday Life*, Michel de Certeau distinguished 'strategies' and 'tactics': a strategy is a procedure for calculating and manipulating power relationships that operates in a certain domain (a city, a scientific institution, an enterprise) that can be isolated from its environment. A strategy has its own, proper place from which it manages relations with what is exterior to it. Tactics are the ways of making do of those who are caught up in spaces that are not their own: the users and consumers of products, the inhabitants of a city, the subjects in a psychological laboratory. Tactics are the ways in which people seize the opportunities such spaces offer and exploit the cracks in the system; they are the tricks and ruses of resistance. 'A tactic is an art of the weak.'[25]

De Certeau offers as a model tactic 'la perruque', the use of the boss's resources for personal ends: printing the football club's newsletter on the company's laser printer, 'borrowing' the workshop's circular saw for renovation work at home, emailing with your girlfriend during office hours. From the point of view of De Certeau's analysis of everyday practices, the resistance that the techniques of social science meet is also tactical. It is the reaction of those who, as the targets or subjects of such techniques, find themselves on a terrain that is organised by a scientific rationality that is not their own (psychologists, after all, are quite emphatic on this point).[26]

Although De Certeau is mostly cited as a theorist of everyday resistance to systems of power, he also extended his analysis of 'the art of practice' to the field of science and the relation between knowing and doing.[27] In science, 'method' systematises thinking as a way of operating – rational procedures are the essence of science. But in the operations of science there are tactics as well. Scientific work does not consist solely in the application of formal methods, but it is also a matter of 'making do' (in French: *bricolage*). This is the '"underside" of scientific activity', 'the stubborn persistence of ancient tricks in the everyday work of … laboratories'.[28] Science, publicly presented as strategic (consisting of rational procedures for the production of facts and theories), is in practice full of tactics as well.[29] Scientists are also workers who do not always follow the rules that are imposed on them and may attempt to subvert for example the system of output measurement by which they are evaluated. But there is more to it than tactics of resistance. More generally science requires not only method but also judgement, or 'tact'. Following Kant's analysis of judgement ('Urteilskraft'), De Certeau writes that it designates 'that which, in scientific work itself, does not depend on the (necessary) application of rules or models'.[30] Judgement, tact, is 'the art of thinking' at the heart of both science and ordinary practice; it is the middle term between theory and praxis. Tact is not a procedure, a method that can be followed. Kant compares tact to the art of tightrope walking: it consists in maintaining a dynamic equilibrium (between imagining and understanding) by continually adjusting to a situation of which one is oneself a part.

In human engineering, these three concepts – strategy, tact and tactics – come together. Human engineering presents itself as a strategy for calculating and manipulating power relationships that promises efficient and effective control over human behaviour. But it also requires judgement, the art of thinking, because it must apply its procedures in concrete situations, balancing the ideals of theory with the constraints of practice. This is the challenge for all engineers: to mediate between theory and practice, to take the techniques that worked so well on paper, or in the lab and in the workshop, and supply whatever know-how is required to make them work in the rough terrain of everyday life.[31] In the case of human engineering, this know-how involves not only judgement of the situation but especially the tact that is needed to prevent and counter resistance. Tact is involved in human engineering because its deployment requires judgement of the situation and because it must deal with the tactics of the people it targets. The tactics of the recalcitrant targets of human engineering require tact on the part of the human engineer. Human engineering is both tact and technology. That, in short, is the scheme that guides my analysis of the historical material that I present in this book.

Overview of the Book

In Chapter 2, I develop the concepts of tact and technology further, drawing on rhetorical theory and science and technology studies. In rhetoric, the question of how to move people is obviously central. Scholars of rhetoric, starting with the sophists in ancient Greece, have long reflected on what persuasion is and how it can be achieved. The concept of *kairos*, variously translated as timing or propriety, invites a further look at tact, in particular the question of whether tact can be formalised. Can the situatedness of persuasion be transcended by formulating it in a finite set of rules, or is it an unavoidably particular, local, individual element in persuasion?

In rhetoric, resistance to persuasion is likewise an important concern. To persuade requires more or less effort depending on the message and the audience. Moreover, rhetorical situations are often implicitly or explicitly polemical. Rhetors may encounter active resistance, to themselves, to their message and arguments and even to the attempt to persuade itself. Some form of concealment is often used to circumvent this resistance.

In fact, concealing persuasion itself has been described as the pinnacle of the art of rhetoric.

Next, I turn to the concept of 'technology'. Although the dream of control over the behaviour of other people is ancient, terms like 'technology' and 'engineering' are of recent vintage. But what in fact is technology? This question becomes particularly urgent when we use terms like 'human engineering' or 'social technology', which are even more recent. Are there two kinds of technology, one social and one not social? In the past few decades, some philosophers, historians and sociologists of technology have started rejecting such distinctions. Simply following them in this rejection is not an option in this study, however, because terms like 'social technology' and 'human engineering' are a part of the history I'm describing. What they referred to, one could say, was at least historically real, in the same way that phlogiston was once real. Whether or not it is sensible to distinguish between 'technology' and 'social technology', people have thought that it was, and from a historical standpoint it is important to understand why they did.

Conceptions of what it means to be human, what a machine is, or control, technology or society, are not static but historically variable. That means I can only use a term like 'human engineering' with the greatest caution, in the knowledge that the meaning of each part of that term has changed over time and is still changing. This is unavoidable in historiography: the vocabulary one uses to write history is to some extent always a part of that history itself. There is no neutral language to work with. The best one can do is to keep one's vocabulary light, be aware of the choices it incorporates and stay sensitive to the discourse of the people one is describing.

The issue of the relation between humanness and engineering was particularly prominent in the controversy over scientific management, which is the subject of Chapter 3. Scientific management, an approach to managing companies and their production processes that was explicitly based on engineering principles, encountered strong opposition over its allegedly inhumane character. In the ensuing controversy, what is human, and how it relates to machines and engineering, was the primary topic of discussion. Psychologists soon began to promote themselves as experts in both human beings and science and engineering and offered their services as 'human engineers' as a complement to natural science and engineering. Management thinker Mary Parker Follett, however, took a fundamental step away from the engineering perspective. She was one of the first to discuss the problem that giving direct orders is not always the best way to get things done because people may resent having to obey orders. It is better, she argued, to present the situation in such a way that the workers discover what needs to be done of their own accord. Follett, in other words, emphasised the importance of tact in management.

Chapter 4 looks at the history of the term 'social technology' in the early twentieth century and addresses the question of why it was surpassed in popularity by 'social engineering'. A key factor was the fact that the concept of 'technology' was itself undergoing change in those years. Whereas it earlier referred to the study of the practical arts, it increasingly meant those arts themselves, particularly the machines and processes in industry and the new consumer appliances like washing machines that industry provided. Now that 'technology' was something material, the term 'social technology' made less sense. 'Social engineering', which denoted an activity, did not share that problem.

In Chapter 5 I turn to one of the masters of tactful control, the public speaking trainer, management consultant and self-help author Dale Carnegie. In his multi-million bestseller *How to Win Friends and Influence People*, he described the art of dealing with people in lists of numbered rules and principles, promising success in business and personal life to those who followed them. But ultimately, Carnegie taught, influence requires more than following rules; it demands a personal transformation. One can only have influence on other people if one is sincerely interested in them.

The concept of 'social technology' returns in Chapter 6, where I take a brief look at the work of the philosopher Karl Popper. In Popper's view, social science is essentially nothing more than social technology, the experimental study of social mechanisms. Popper formulated this idea in reaction to the totalitarian utopianism of fascists and communists and the holistic tendencies in social science. Society cannot be understood as a whole, nor must one attempt to transform it in its entirety. Only by careful, cautious, piecemeal experimentation and engineering can society be understood and improved while avoiding dictatorship and oppression. Moreover, Popper was emphatically against any attempt to engineer individual behaviour and opinions: psychological engineering would be both unscientific and undemocratic, which, in Popper's philosophy, amounted to much the same thing.

Around the same time, in the United States, social psychologist Kurt Lewin and his students did develop a psychological technology in the service of democracy. Their theories, experiments and techniques, which I describe in Chapter 7, focused on the small group. Social order could be created democratically, they believed, with techniques of leadership that stimulated the group to make its own decisions. Such leadership requires considerable tact in order to negotiate the tension between the directive actions of the leader and the responsible self-government that is the aim.

I argue that this tact remained to some extent embodied in experienced leaders and was not fully articulated in scripts derived from theory. Tact had to be acquired by observing and doing (unless one already had it) and could not fully be read from a script. To some extent, each situation required a new translation from theory to action.

Control and resistance were the central issue in the brainwashing scare of the 1950s, the topic of Chapter 8. Apparently sincere but false confessions made by Soviet prisoners, American prisoners of war in Korea and Chinese victims of 'thought reform' sparked the fear that the Communists had discovered an effective and efficient way of controlling mind and behaviour and were using it for the purpose of spreading Communism. The CIA paid scores of scientists of various disciplines, with or without their knowledge, to investigate the Communist methods and to develop mind control techniques for its own use. More than a decade of research did not yield a mental miracle weapon but rather confirmed the importance of tact in any attempt to influence others, even if physical violence is used to break resistance.

In Chapter 9 I describe the current crisis in psychology. The discipline is gripped by doubts about the quality of common research practices and the trustworthiness of some of the claims made in research reports. This crisis of confidence was spurred by several cases of fraud and by studies showing the prevalence of so-called 'questionable research practices' but perhaps even more by failures to replicate the results of a number of high-profile psychological experiments, chief among them the priming studies of John Bargh and Ap Dijksterhuis. The controversy around replication and priming allows me to follow tact and technology back into the lab and ask to what extent the experiment is itself a technology. Finally, Chapter 10 concludes the book with a brief review of the dynamic of control and resistance that I described in these histories.

2 Tact and Technology

Before psychologists got involved, persuasion was the expertise of rhetoricians. The rhetorical tradition offers a rich storehouse of reflections on the work of influencing people's mind and behaviour, and I will draw on it to shed light on current practices of human engineering. Consider, to begin with, Gorgias of Leontini, one of the itinerant teachers of public speaking known as the sophists. The way that Gorgias described the power of speech to move people is reminiscent of the claims of current human engineers. His most famous text is the *Encomium of Helen*, in which he argued for the innocence of a woman despised by every Greek.[1] Helen, according to Greek mythology, was the most beautiful woman on earth. She was the wife of King Menelaus of Sparta, but she eloped with the Trojan prince Paris, an insult that needed to be avenged and led to the ten-year-long bloodshed of the Trojan War. Consequently, Helen was the most reviled woman in Greek history: a traitor and an adulteress in a culture that considered women inferior to men in the first place.

To persuade his audience that Helen was, nonetheless, not to blame, Gorgias argues that her actions have four possible causes: the will of the gods, physical force, speech or love. That it is impossible to resist the gods (or chance or fate) is obvious, and the same goes for physical force. If she had been seized by force, clearly the man who seized her did wrong, not Helen. More surprisingly, Gorgias offers similar arguments for love and speech. When people see frightful things, their fear can be so strong as to 'extinguish and expel thought'[2] and make them flee in utter panic; the effect of pleasing sights, such as Paris's body, can be equally strong, and '[i]f love is a god with a god's power, how would the weaker be able to repel and resist it?'[3] Of the four possible causes, Gorgias devotes most attention to *logos*. Speech, he stresses, can be a causal force exactly like physical violence, the will of the gods or the emotion of fear. Persuasive speech in particular compels, 'it moulds the mind in the way it wishes', in a manner that is entirely analogous to the action of physical causes.[4] 'The power of speech bears the same relation to the ordering of the mind as the ordering of drugs bears to the constitution of bodies.'[5]

Gorgias ends his speech by saying that he wrote it 'as an encomium of Helen and an amusement for myself'.[6] It probably also served as a demonstration piece, advertising his rhetorical skills. A defence of the treacherous Helen must have been a particularly impressive trick to pull off. Nevertheless, the speech is more than a bit of fun to impress potential clients. Douglas MacDowell notes (in the introduction to his translation of the *Encomium*) that everything we know of Gorgias points to the fact that he really considered speech to be a very strong, if not irresistible, way of moving people. As Gorgias put it in the *Encomium*: 'Speech is a powerful ruler. Its substance is minute and invisible, but its achievements are superhuman.'[7] The *Encomium* is cleverly reflexive: it demonstrates Gorgias's skill by reversing his audience's opinion of Helen, with the argument that persuasive speech (whether of Helen's abductor or of Gorgias himself) is a force that neither Helen nor anyone else could possibly resist. It is at once a demonstration of Gorgias's powers and a speech about the power of the word.

There are parallels between the dream of behavioural control as expressed by current human engineers and Gorgias's praise of the superhuman achievements of speech. Arguably, both serve to advertise the capabilities of a group of professionals: neuromarketers, public speaking trainers, sophists. But there are other similarities as well. Like his modern counterparts, Gorgias pictured persuasion as a process that is analogous to the action of physical agents. Persuasive speech is a causal force that moves people like the wind moves a ship. It is not the rationality of the words that persuades the audience: reason or thought have nothing to do with it. It is noteworthy that Gorgias compares the power of speech to that of emotions like fear, which can be so strong as to 'extinguish and expel thought'. Likewise, speech is powerful precisely when it bypasses thought. The basis of persuasion lies not in knowledge, as Plato thought, but in the emotional response of the audience to the words of the speaker.[8] To persuade, speech must trigger the emotions, not reason.

Modern human engineers would agree. They too downplay the role of consciousness and rationality in human behaviour, instead pointing to the importance of unconscious influences and emotional reactions. According to De Munnik, all our decisions are emotional, and none of our decisions has anything to do with the ratio (other than that the ratio may have influenced our emotions beforehand). Recent research, he adds, points in the direction of 100 per cent of our decisions being taken unconsciously and automatically.

But Gorgias is known not only for his praise of the power of persuasive speech but also for his emphasis on the importance of the situation. To be persuasive, speakers need to adapt their words to the situation at hand: to the location, the circumstances, the particular audience that is being addressed. What works in one place at one time may not work in another. A speaker needs timing and a sense of what his audience will consider right and proper. Gorgias and other Greek authors on rhetoric referred to this quality of the rhetorical situation as *kairos*, a concept that has been called 'the cornerstone of rhetoric in the Golden Age of Greece'.[9] *Kairos* is a concept of time – not the time of clocks (*chronos*) but rather the qualitative aspect of time: timeliness, occasion, time that is good or bad. To be persuasive, a speaker needs to adapt to the *kairos* of the situation, to the opportunities it offers and the constraints it sets. Rhetorical scholarship about *kairos* offers a way to further develop and refine the concept of tact.

Kairos has been interpreted in different ways.[10] One view is that it demands that speakers accommodate their words and actions to convention and to the expectations of their audience. *Kairos* is then a matter of decorum. As Cicero put it: 'The universal rule, in oratory as in life, is to consider propriety.'[11] This is a moderate, somewhat conservative view of tact as a concern with what is proper and expected. A more radical view is that *kairos* points not to the importance of conventions but rather to the unexpected in rhetorical situations. In this view, persuasion always contains an element of contingency and improvisation. Despite the many rhetorical handbooks, there can be no fail-safe recipe for persuasion. Decorum may be an important consideration for speakers, but above all they must be creative.

These two interpretations of *kairos* imply different views on tact and the reach of what Pacelle van Goethem's website calls 'the laws of persuasion'. If *kairos* points only to the importance of propriety, of adapting a discourse to the particular situation and audience, then the rules and principles will have to include a situational component, but persuasion is essentially a lawful process. As long as the characteristics of situation and audience are known, the process of persuasion can be fully described in laws and the words and actions of the speaker fully prescribed. If persuasion is radically contingent, however, as in the second interpretation of *kairos*, the actual situation in which persuasion occurs will always be beyond the reach of laws. They can function only as principles or heuristics that a speaker must adapt to the contingencies of the place, the moment and the people present. When it comes down to it, getting people to think or do what you want can never be fully prescribed by a manual. Tact is never entirely encompassed by technology; it always remains as an irreducible, situational demand.

Michael Billig, in *Arguing and Thinking*, a plea for a social psychology that is informed by the rhetorical tradition, illustrates the first approach to rhetoric with the views of Plato. In *Gorgias*, a dialogue between Socrates and the sophist, Plato mounted a blistering attack on Gorgias and other sophists (they were only in it for the money) and their so-called art, which was nothing more than a collection of tricks. A true art or *technè* is grounded in rational knowledge. In another dialogue, the *Phaedrus*, Plato spelled out his vision of a future art of rhetoric: it will be based on knowledge of 'the different kinds of speech and different kinds of soul, and their different conditions', and it will 'enumerate all causes that act, and suiting kind by kind, will show what sort of soul is of necessity persuaded, or not persuaded, by what sort of speech, and for what reason, in either case'.[12] Plato, in other words, was envisioning a rhetoric that fulfils the dream of efficient and effective control over human behaviour. It will be a technology that encompasses tact by fully specifying the situational aspects of persuasion and how they should be handled: it formalises tact as a finite number of if-then rules.

Billig does not believe there will ever be such a rhetoric. Human behaviour and the society we live in are simply too diverse and changeable to allow the kind of knowledge that that requires. There is always something novel in a rhetorical situation, and therefore '[r]hetoric can never be a simple matter of mechanically classifying the particular case as being an instance where this or that tactic will be persuasive'.[13] Billig calls this 'Quintilian's uncertainty principle', after the Roman rhetorical scholar: 'we can never capture the infinite variety of human affairs in a finite system of psychological laws'.[14] There is always an element of indeterminacy about persuasion and consequently always a 'need for hunches or judgement'.[15] A technology of persuasion always needs to be complemented with tact, and being tactful cannot be specified with a list of rules.

Hidden Artifice

As we have seen, resistance is a complicating factor in human engineering. According to Van Goethem, it even has its own brain area – the amygdala – prone to stress at the least provocation. In the rhetorical tradition it has long been recognised that the very act of persuasion itself can be provocative. People do not like being told what to do or think but prefer to make up their own mind. They're best persuaded if they're not aware of how they're being persuaded or, better yet, of the fact that they're being persuaded. Therefore rhetorical handbooks often advised not to be too obvious about one's intent and techniques. As Carolyn Miller noted, in the history of rhetoric an impulse towards self-aggrandisement has always been accompanied by a predilection for self-denial.[16] Quintilian considered the orator's art to be best expressed in 'the concealment of its existence'.[17] Lysias was highly admired for the way he concealed the rhetorical techniques in the speeches he wrote for others, including the fact that they had been written by a professional 'logographer': they appeared to be the words of the speaker himself, speaking off the cuff.

A good way to start a speech, according to ancient Greek handbooks, is to deny any rhetorical skill.[18] Thus, Socrates began his defence in the trial against him with an indignant denial that he had a way with words, as the prosecutors had claimed. 'I thought it was particularly brazen of them to tell you this without a blush, since they must know that they will soon be effectively confuted, when it becomes clear that I haven't the slightest skill as a speaker – unless of course they mean by a skilful speaker one who speaks the truth.'[19] Instead of flowery language, what his fellow citizens will hear from him 'will be a straightforward speech in the first words that occur to me'.[20] The tactic remains popular, surviving, for example, in the common opening phrase 'unaccustomed as I am to public speaking. . .'.[21]

As Miller puts it, '[r]hetoric, it seems, must deny itself to succeed'.[22] Moreover, the self-concealment of rhetoric is always a double deception: not only is it advisable to hide the art, it is essential 'to *appear* not to conceal'.[23] As Dyonisius wrote of Lysias's skill: 'it is in the very illusion of not having been composed with masterly skill that the mastery lies'.[24] Thus, creating the appearance of spontaneity and artlessness seems to be a crucial rhetorical skill, and one that itself of course must be concealed at all cost. Modern persuaders are aware of this. Van Goethem's best-known seminar is called 'Invisible Persuasion'.[25] Similarly, a recent book divulging 'the psychological tactics and tricks to win the game' is titled *Covert Persuasion.*[26]

MasterFlirt, a manual for aspiring pickup artists, is a perfect example of the combination of self-aggrandisement and self-denial that Miller noted in the history of rhetoric. The book opens with congratulations to the reader: he has just put his first step on the way to becoming a MasterFlirt, and the results will surpass his wildest dreams. But then follows a stern warning to the student: 'Don't tell any woman you're studying *MasterFlirt*. You will see that many things that we discuss give you access to a woman's unconscious. As soon as she notices that, she will immediately put up a blockade.'[27] Only veiled techniques will give you access to the inner recesses of a woman's mind.

Deception holds a central place in psychology. In his history of the relation between psychology and commerce, Michael Pettit has described how the discipline, at least in the United States, has been fascinated ever since its early days by the deceivable as well as the deceitful self.[28] The psychological subject was seen as vulnerable to suggestion, to illusions, to the tricks of con men and the wiles of advertisers, but at the same time the subject could not be trusted either, being unwilling or unable to divulge the truth about herself. Psychologists positioned themselves as guardians of reason and truth, exposing illusions and detecting deception.[29] But in the pursuit of truth about human nature psychologists, social psychologists in particular, consider it necessary to engage in deception themselves and to remain unobserved as observers. To make sure experimental subjects are honest and natural and to prevent their resistance it is necessary to keep them 'naive' and mislead them about the purpose of the experiment.[30] 'The use of trickery, guile, and deceit to elicit a supposedly natural response from experimental subjects became a major component of twentieth-century psychology.'[31]

Outside the laboratory, too, the techniques of psychology operate best when they're veiled. When B.F. Skinner's techniques of operant conditioning were transferred from the laboratory to real-life contexts such as prisons and mental institutions, many people objected strongly to their 'inhuman' and 'mechanistic' character. Both Alexandra Rutherford and Karen Baistow have shown, however, that Skinnerian behaviour technology met with great success when it was shorn of its radical behaviourist philosophy and offered as a set of techniques for self-control. Skinner's theory of environmental determinism, according to which there is no such thing as self-control, proved a hindrance in the spread of his techniques, but embedded in ideas and practices of empowerment his technology of behaviour modification proved popular and successful.[32]

The resistance that human engineering has to contend with is thus partly directed at the fact that it *is* engineering, and the tact it requires is partly a matter of concealing its own mechanisms. In this book I will explore several variations of this paradoxical relation between tact and technology. It's also important, however, to note the contingency of this paradox: whether artifice needs to be concealed depends on a number of factors that can vary. Carolyn Miller describes two of them. Although the tension between artifice and concealment has a long history in rhetoric, Miller points out that it rests on two assumptions: that human relations are adversarial and that language is mimetic. The first implies that artifice is not persuasive because it triggers suspicion. Greek and Roman rhetorical handbooks pictured a highly adversarial, forensic view of rhetoric, with the court of law as its primary arena, shot through with hostility and resistance. In such a setting, any hint of artifice would raise the hackles of the audience. It is better to be, or appear to be, sincere and spontaneous, 'natural' rather than 'artificial'. In an adversarial situation, a natural style creates trust. It relaxes people, as Van Goethem would say.

This preference for sincerity and spontaneity also derives from the idea that language is mimetic, that it can mirror reality, provided it is not warped or obstructed by artifice and deception. If we just let it flow unimpeded, it becomes transparent. In its natural, uncorrupted state, language is a neutral vehicle for transmitting the truth about the world. Therefore the more spontaneous and sincere we seem to be, the more we appear to speak the truth. When these two assumptions pertain, when the rhetorical situation is adversarial and language is expected to be a mirror of the world, then concealing persuasion and its tricks is the essence of the art of rhetoric. However, this is not always the case: sometimes an audience cooperates with the speaker in the rhetorical event and expects the speaker to use language to create reality rather than merely mirror it. A funeral oration is a good example of such a situation. Typically, the mourners do not expect the speaker to be brutally honest about the deceased, nor is spontaneity or a natural style particularly appreciated. On the contrary, rhetorical flourishes may be acceptable even when they're obvious and over the top. The best funeral speeches do not hide their artifice but openly use words to bring the audience to tears or provide welcome relief with a humorous anecdote. The role of the speaker is not to portray the deceased as faithfully as possible but to provide an expression for the anguish of the bereaved. A good funeral speech does not merely describe sadness; it creates an event of collective grieving. Its language is not mimetic but performative. It doesn't portray; it enacts.

What is true of the funeral goes for all situations that call for well-crafted praise, including weddings and award ceremonies. All are occasions for cooperation and performance. Aristotle called this branch of rhetoric the 'epideictic' and distinguished it from the forensic, adversarial rhetoric of the courtroom. The third category he delineated was deliberative rhetoric, typical of politics. Here, although the discourse is again argumentative, the goal is not the victory of this or that party (at least, it shouldn't be) but the good of all. Ideally, deliberation is a cooperative process conducted by equal partners, pitting arguments against each other to discover the best course of action. Thus, classical rhetorical scholars recognised that different situations call for different styles of discourse. The assumptions that underpin the rhetoric of concealed artifice according to Miller are pervasive, but they're not ubiquitous. Persuasion can be, but is not always, a matter of more or less deceptive manipulation. (Neither, one would hope, is seduction always, or even usually, a competition between men with women as their unwitting prey and prize.)

Technology

The relation between tact and technology in human engineering raises the question of whether it is different from engineering proper. Does tact mark these technologies as distinct from other technologies? Is there something specifically human about human engineering? This question, in turn, raises another: what in fact is technology or engineering? In a famous essay the philosopher and anthropologist of science Bruno Latour proposed an answer to both these questions. In '*The Prince* for Machines as Well as for Machinations' Latour compares the notoriously cynical ploys described by Niccolò Machiavelli (the 'machinations') with modern technology (the 'machines').[33] In his unending quest to maintain and expand his power, the prince of Machiavelli's day needed support. Although he could make use of some 'non-human allies' – Machiavelli discusses fortifications and weapons – mostly he had to depend on the loyalty of other men such as soldiers, other princes, the populace and the Pope. The prince's world is primarily one of human relations that constantly need to be managed. Thus, 'Machiavelli builds his plots by keeping men in check through the handling of other men who are in turn kept in line by other men. . . . To constantly repair the decaying social order, social forces are, if not the only, at least the main resources.'[34]

Modern princes have many more allies at their disposal. Today, power is wielded not only by managing human relations with subtle machinations but also by manipulating natural forces with clever machines. A modern prince has the 'obstinacy, the cunning, the strength of electrons, microbes, atoms, computers, missiles' at his disposal.[35] It would be a mistake, Latour warns, to see these as two separate realms: society and technology, the one human and the other natural. If we follow the powerful, we see that they combine and tie together all sorts of allies. They are, in the felicitous phrase of John Law, 'heterogeneous engineers', who weave together human and natural resources into 'socio-technical imbroglios'.[36] It makes no sense to keep distinguishing nature and society: with each engineering innovation, the heterogeneity increases. Technology is not a matter of harnessing the forces of 'nature' to be used by 'society'; it is the combining of human and non-human forces into new arrangements that expand the power of the princes.

What distinguishes modern rulers from their Renaissance counterparts is that they have many more potential allies they can mobilise. Whereas Machiavelli's prince had little more than other men to rely on and love and fear to tie them to him, modern princes can make use of a long, mixed list of human and non-human allies. We mustn't attempt to separate machines from machinations – a machine is a machination but with a more diverse composition. Human engineering is not a different, social kind of technology. All technology is social. The key question is not 'is this a human, social ally, or is it a natural one', but how strong are the associations between the elements in this socio-technical arrangement?

Latour's take on the question 'what is technology', and that of allied thinkers such as John Law, Michel Callon, and Madeleine Akrich, has been very influential. In their wake, many researchers have explored the heterogeneous engineering that weaves together human and non-human elements and forces into socio-technical assemblies. The attraction of this approach lies not only in its view of how technology develops but perhaps even more in its perspective on the role of technology in society. It reveals technology as more than a collection of tools deployed for ends that are decided by 'social processes': our society is to an important extent a technological creation, in which technologies such as cars, mobile phones, birth control pills, electric lighting, shipping containers and much more not only facilitate but also create ways of relating between people. Our society is not one of human beings among themselves, using tools to master nature; the tools are an integral part of the 'collective' (a term that Latour prefers over 'society') and increasingly of the human beings themselves, as they become physically and mentally entangled with their laptops, smart phones, pharmaceuticals and other technologies.

Although Latour warns against distinguishing machines and machinations – since modern, heterogeneous engineering consists in weaving together human and non-human forces – there remains for him an important difference, one that makes his work particularly relevant for this study: human bonds are soft; non-human bonds are hard. As he wrote in 1988, 'the Principates had to rely on ties such as passions, fears, love and ambitions, as soft as the bodies they had to attach'.[37] The non-human allies of the modern princes are made of sterner stuff: '[t]o the piling of tender human bonds will be added the piling of harder non-human bonds'.[38] (For example, Ukraine is still firmly tied to Russia by a gas pipeline, although the human bonds between the two countries have partly disintegrated.) In later texts, Latour returned to this issue, continuing to emphasise the hardness, the durability of non-human associations: 'It is always "things" – and I now mean this last word literally – which, in practice, lend their "steely" quality to the hapless "society".'[39] His favourite contrast is with baboons, whose society is constantly teetering on the brink of chaos. Having only his 'basic social skills', the alpha male has to resume repairing the associations in his group each morning. Our own heterogeneous, socio-technical collectives are made extensive and durable by the inclusion of objects, artefacts and devices. It is the 'capacity of artefacts to construct, literally and not metaphorically, social order' that gives our collectives 'durability, expansion, scale, mobility'.[40] The moral and legal obligation to wear seat belts is far better enforced with material means such as annoying lights and sounds when you forget to put them on than with traffic safety campaigns.[41]

This emphasis on objects has proven to be a fertile approach to studying modern society, with Latour's work inspiring a materialist turn in social theory.[42] As revolutionary as Latour's proposal was, however, the idea that human bonds are soft and human allies unreliable is rather conventional, echoing as it does the traditional distinction of the hard and the soft sciences. More to the point, to ascribe certain characteristics to human beings and their associations in this way flies in the face of Latour's own constructivist perspective. In his approach to the study of scientific processes, he usually avoids assuming facts about the world, facts that are waiting to be discovered by science. Facts are made, not in the sense that they're made up out of thin air but in the sense that they're constructed in the interactions that make up the scientific process. Scientific statements are not representations of a world independent of us; they are the endpoints of a long line of associations between a heterogeneous collection of human and non-human actors, including scientists and the material they work with. To understand the scientific process we must study how these associations come to be.

A crucial principle of Latour's method is therefore that we shouldn't 'decide in advance on a list of actors and possible actions'[43] because 'we do not know in advance what the world is made of'.[44] To assume at the outset that human beings are fickle doesn't square with this principle. It would be more in keeping with Latour's methodology to treat the characteristics of human beings as the *outcome* of processes of association between heterogeneous actors. Such processes include the development of techniques of human engineering. People are not objectively, essentially unreliable: they are strong or weak in association with particular actors, and we can study the history of human engineering as the history of such associations from which human bonds of a certain strength emerge. In this study, we do not 'know in advance' what people are made of.

Latour's work is also relevant for my purposes because resistance is a central term in it. In his programmatic (and metaphysical) text 'Irreductions', Latour formulated a minimalist point of departure: 'there are only trials of strength'.[45] We do not know in advance what the world is made of, except that it consists of a heterogeneous variety of actors (or 'actants', in Latour's terminology) that are involved in trials of strength. In this perspective, real is what resists trials. A scientific experiment is such a trial, and it results in a particular articulation of the reality of the actants involved in it. The spectacular success of the natural sciences is not due to their control over nature; on the contrary, the objectivity of the natural sciences lies in their ability to let 'objects object to the utterances we make about them'.[46] The natural sciences thrive by seeking out the resistance of the objects they study, not by mastering them.

According to Latour the social sciences have misunderstood this, and in their eagerness to emulate the natural sciences they have made control over their objects of study their central methodological principle.[47] This seems to work well because people are prone to docility; they are easy to bring into line. As Milgram's famous obedience study showed, an experimenter with a white lab coat and a clipboard is all that is needed to turn them into obedient killers. Compliance comes easily to them. Non-humans, on the other hand, tend to resist. 'Natural objects are naturally *recalcitrant*; the last thing that one scientist will say about them is that they are fully masterable. On the contrary, they always resist and make a shambles of our pretensions to control.'[48] It is that resistance that produces new ways of articulating their reality. Instead of seeking control, social scientists should also allow and encourage their subjects to be disobedient: 'it is precisely when the objects of study are interested, active, disobedient, fully involved in what is said about themselves by others, that it sometimes happens that a field of a social science begins imitating for good the surprising novelties of some of the best natural sciences'.[49]

Latour's claim that people are docile and easy to control by social scientists seems to go counter to his insistence, in other places, that people are fickle and hard to bring into line by princes. More importantly, I believe Latour is again violating his own methodological principles here. Rather than ascribing to human beings certain characteristics (docility, fickleness) in advance, we should consider them as the (provisional) outcome of ongoing trials of strength. The social sciences, psychology in particular, are in the business of 'trying' people's resistance: when are people docile or recalcitrant? Under what circumstances will they comply, and when will they resist? Control and resistance are important topics of psychological research, and it seems odd to assume that the answers can already be given, when the task that historians and sociologists of science usually set themselves is to explore how the answers are arrived at. It is true that the social sciences, psychology in particular, often emphasise the manipulability of human beings. This, after all, is the core of the dream of efficient and effective control over human behaviour. But we can also see the traces of the resistance of the objects of study, particularly in the tact that is required when dealing with them. By reading human engineering against the grain, so to speak, we can bring that resistance to light.

Return of *The Prince*

As a first step towards a more empirical approach to human resistance and the reliability of human bonds, consider again *The Prince* and its advice on how people can be manipulated. Machiavelli's view of human beings and how to handle them is more complex than Latour's two-dimensional interpretation of it. It is true that Machiavelli warns the prince not to depend too much on others because 'the people are fickle; it is easy to persuade them about something, but difficult to keep them persuaded'.[50] People are 'ungrateful, fickle, feigners and dissemblers',[51] they love and support you when it is to their benefit, but when you actually need them they are nowhere to be seen. Yet they are also creatures of habit. For this reason it is easier to hold on to a hereditary principality than to a new principality, but also in cases like the latter, a skilful ruler, who has made sure that his citizens depend on him and his government, can trust them to 'always be loyal to him'.[52] When cementing his alliances and outsmarting his opponents, the prince can rely on many facts of human nature. For example, 'men are so constituted that they are as much bound by the benefits they confer as by those they receive'.[53] Hence, they will support you when they feel you are indebted to them.

Thus, although people are fickle, there are nonetheless many regularities in their behaviour that a shrewd ruler can and must count on. Machiavelli's famous question whether it is better to be loved than feared or vice versa has a definite answer: if you have to choose, make sure you are feared. 'Because love is sustained by a bond of gratitude which, because men are excessively self-interested, is broken whenever they see a chance to benefit themselves. But fear is sustained by a dread of punishment that is always effective.'[54] Never incur the hatred of the people, however, because a populace that hates you is certain to bring you down, no matter the strength of the fortress that you're hiding in.

The skill of a ruler lies not only in his knowledge of human nature and how to exploit it but also in how he deals with the vagaries of fortune. Machiavelli's concept of 'fortune' not only covers good and bad luck but more generally the circumstances one finds oneself in, circumstances that are not always under one's control. A skilful ruler knows both how to adapt when fortune is against him and how to exploit the opportunities that may arise. In other words, he needs to be in tune with the *kairos* of the situation, with the 'occasione' (as Machiavelli put it).[55] 'Occasione' also explains Machiavelli's notorious cynicism because the timeless dictates of morality are useless in the face of ever-changing circumstances. A ruler should be ready to 'vary his conduct as the winds of fortune and the circumstances constrain him' and 'be capable of entering upon the path of wrong-doing' if need be.[56]

Thus, while Machiavelli builds on the laws of human nature to formulate advice that is sure to help you to power, he balances this universalism with an equal emphasis on the importance of the situation, the specific people involved and the skill required to adapt one's strategy to them. The word may seem ill-suited given Machiavelli's uncompromising realism, but the ruler needed tact. In line with this emphasis on 'occasione', much of *The Prince* consists of descriptions of situations that historical and recent rulers found themselves in and of how they acted upon the necessities and opportunities inherent in the occasion. Great men like Cyrus and Theseus were fortunate to be given an opportunity to realise their goals, had the perspicuity to see that opportunity and had the abilities to exploit it. To acquire a sense of 'occasione' one has to do more than memorise lists of general principles; one must study actual occasions in which virtuous leaders displayed their superior judgement. A sense of the particular can be developed only by studying particulars and how they illustrate general principles.

When considering Machiavelli's advice as a model for machinations in general, as Latour does, it is crucial to keep in mind the specific problem to which it was addressed: how absolute rulers of principalities in Renaissance Italy could gain and maintain power. That is a context that differs from our own in several key ways. One of these differences lies in the role of force. For Machiavelli, violence is always an option, and instilling fear is one of the main tactics. There are many ways to create alliances – handing out gifts and bribes, bestowing honours, preferential treatment – but the possibility of using force almost always hovers in the background as an alternative. Machiavelli presents the options in no uncertain terms: 'men must either be won over or else destroyed', adding that his model ruler, Cesare Borgia, had understood this well.[57] Machiavelli frowns upon those who seek a middle way in leniency and light punishment. This will lead only to problems. '[M]en should either be caressed or crushed; because they can avenge slight injuries, but not those that are severe.'[58] Force is an instrument of power just like money or laws, but it has to be used well. Cruelty must be applied swiftly and thoroughly and be followed by beneficial measures in order to win back the loyalty of the citizens. Continuing cruelty is sure to incur hatred and lead to one's downfall.

Secondly, and related to this first difference, it is important to note that 'the people' play a role in Machiavelli's considerations only inasmuch as princes require the support of notable citizens and of the populace to maintain power. This support is crucial, and 'wise rulers have always been very careful not to exasperate the nobles and also to satisfy the people and keep them contented'.[59] To this end, the prince must make sure that he avoids incurring the people's hatred and contempt. Hatred is fairly simple to avoid: all he needs to do is keep his hands of their property and their women. Avoiding contempt requires more effort. To keep the respect and admiration of the people, the ruler's actions must be forceful, courageous and grand so that he will not be considered 'inconstant, frivolous, effeminate, pusillanimous, and irresolute'.[60] It is not necessary, and often ill-advised, that he abide by traditional virtues such as honesty and generosity as long as he is able to create the appearance of being honest, generous and so on. A truly virtuous ruler (that is, one who has what it takes to hold on to power) combines the strength of the lion with the cunning and deceptiveness of the fox, to which Machiavelli adds that 'foxiness should be well concealed'.[61]

What is missing in Machiavelli's discussion of the relation of the prince with the people is the issue of governance. Statecraft in *The Prince* is primarily a matter of control over territories, not of governing a state and maintaining social order. A prince is a warrior first and foremost. 'A ruler should have no other objective and no other concerns, nor occupy himself with anything else except war and its methods and practices.'[62] Machiavelli does note that 'the main foundations of all states . . . are good laws and good armies', but '[s]ince it is impossible to have good laws if good arms are lacking, and if there are good arms, there must also be good laws',[63] he doesn't discuss laws any further. Instead, Machiavelli devotes an extensive discussion to the different types of armies, pressing the point that a ruler must have his own army and not be dependent on mercenaries or the forces of an ally.

Machiavelli's perspective in *The Prince* is constrained not only by the time he lived in, a time in which power depended mostly on mobilising human allies because non-humans were not as integral a part of the collective as they are now. It is also limited by the figure he focuses on, one whose goals centre on gaining and maintaining absolute power, for which he has means at his disposal that include various kinds of violence such as torture and assassination. To generalise this perspective to cover all kinds of machinations, as Latour seems to do, is problematic. That is not to say that absolute power and torture are now things of the past – Chapter 8 will discuss a rhetorical practice in which power difference and violence are essential. But it is important to widen the view to include modern forms of governance, the technologies that have emerged from them, and the kind of resistance that they have to contend with.

Government

According to the influential analysis of Michel Foucault, the current, Western mode of government subjects people less by force than by knowledge.[64] Social order is no longer maintained by dispensing spectacular forms of punishment, such as public beheadings, in the name of a highly visible sovereign but by the operation of strategies and techniques that make individuals and their behaviour calculable and manageable. Instead of the sovereign, it is now the subjects that are visible, and instead of violent punishment and repression, we have more subtle but no less powerful techniques for the regulation of people's actions. Foucault showed that this type of governance is connected with the emergence of the scientific disciplines that deal with human conduct. The knowledge that these 'psy disciplines' have produced is more than the collection of simple facts of human nature that Machiavelli's prince had to rely on. It comprises a range of 'intellectual technologies, assemblages of ways of seeing and diagnosing, techniques of calculation and judgment'.[65] Secondly, the psy disciplines contribute to a myriad of techniques for the 'conduct of conduct', for governing in the broad sense of steering, directing, regulating and influencing people's actions.

Thus, governing is no longer a matter of either 'caressing or crushing' people, as Machiavelli advised, but proceeds by examination and administration, by education, correction and therapy. It is also no longer exclusively the business of the prince or 'the government': a range of experts with training in the psy sciences is at work in society, occupied with the 'calculated transformation of human conduct'.[66] Since Machiavelli's days, to the prince's machinations have been added a multitude of subtle and intricate ways of governing, deployed by legions of professionals in a diversity of settings, from schools and companies to psychiatric clinics and prisons. Latour's emphasis on material devices as the backbone of current society risks missing the crucial role of the psy disciplines in its governing.

A key aspect of governance in Western, liberal democracies is that it tends to operate via the autonomy of the subject. As Rose famously put it, 'the modern liberal self is obliged to be free', is expected to take responsibility for his own health, happiness and career.[67] Government is not so much a matter of top-down command and control as of encouraging and facilitating self-government, of allowing subjects to make their own choices while also expecting them to make the right choices. The term that Foucault used for this 'contact between the technologies of domination of others and those of the self' is 'governmentality'.[68] Psy expertise is crucial in governmentality, offering, among others, pedagogical techniques to produce such self-governing individuals, therapies to restore their autonomy, and forms of management to regulate their conduct. An example of the latter is the management by dialogue described by Mads Karlsen and Kaspar Villadsen.[69]

Karlsen and Villadsen note a shift in recent management thinking away from hierarchical relations and top-down control towards a more egalitarian style that encourages the initiative, creativity and self-supervision of employees. Techniques of dialogue are an important instrument in such a form of management. Their objective is to get employees to speak authentically, to make them speak their own mind and freely give their opinion. This may seem like a reversal of roles between 'the one who used to speak' (the superior) and the employee, but Karlsen and Villadsen note that the superior's contributions to the dialogue, including the silences, are strategic.[70] They structure and direct the employee's speech and generally make the employee take responsibility for her own words. By reflecting and summarising certain statements rather than others, for instance, the listener can fixate the employee on a particular position and make her acknowledge it as her own. There is a framework that structures the available positions (one of 'needs' or 'dreams', for example), but within that framework the employee is free to speak or, rather, must freely speak.

In this way, the dialogue offers a way to navigate what Karlsen and Villadsen see as fundamental paradoxes in liberal forms of government: 'how does one foster and govern individuals, who are fundamentally self-governing? How does one bring the individual to recognise a particular speech as his or her own? How does one turn the individual into an author of statements, which are produced, selected and arranged through discursive ordering procedures?'[71] In other words, such techniques of dialogue facilitate the production of authenticity and make it part of the government of the work situation. They elicit, within a particular framework, spontaneous, authentic statements with which the employee takes upon herself part of the task of management.

A second example of such a technique of eliciting spontaneous, authentic statements is the focus group.[72] In a focus group, the conversation of a group of people gathered together in a room (and often observed through a one-way mirror) is moderated in order to produce individual statements of opinions on a particular subject. It is a tool that is applied in many domains but is especially appreciated in marketing. Javier Lezaun has described the central role of the facilitator or moderator in managing two paradoxical aspects of focus groups. Firstly, the moderator must resolve the tension implied in using a group process to produce individual opinions. A moderator 'must make the conversation conducive to the expression of private and idiosyncratic views, while preventing the focus group from rising to the status of a "collective"'.[73] Secondly, a central concern of the moderator is the tension between artificiality and naturalness.[74] The artificiality of the focus group situation, embodied in the presence of the moderator, must engender a natural conversation in which authentic opinions are spontaneously expressed.

Thus, 'the fundamental problem for moderators is how to turn the research subjects away from the experimental features of the setting'.[75] The participants are of course aware they are there for a reason, try to figure out what it is and often either resist it ('sabotage', as it's called) or try to be helpful, neither of which produces authentic opinions. The moderator must carefully manage the group to express 'non-directive' opinions but must do so without drawing attention to his tactics. '[N]on-direction needs to be actively engineered into the behaviour of the moderator and into the responses he elicits from the research subjects.'[76] 'The goal of the moderator is to benevolently (forcefully, yet imperceptibly) lead the focus group to a useful outcome (of which their subjects are ignorant).'[77] Lezaun notes that a picture emerges from the manuals and handbooks of the skilful moderator as 'the chameleon-like master of an arcane art of conversation'.[78] Another way of expressing the moderator's skill is that it involves tact as well as technique. A moderator needs to be sensitive to the *kairos* of the situation, be aware of the particular personalities present and his own influence on them and be able to modulate that influence while at the same time hiding his artifice. The focus group is a good example of the combination of tact and technology.

In the light of such analyses of the tools of modern governing, Latour's emphasis on the fickleness of human beings and the softness of human bonds is a bit off the mark. The most prominent feature of people that is addressed by techniques of governance is not so much their lack of firmness but rather their autonomy, their own will, and their own authentic opinions and desires. This autonomy and this individuality are required of them, in the focus group situation as well as in society generally, and must be made productive. The challenge then is not so much to create dependable bonds with inherently unreliable human allies but to manage the paradoxical tasks of regulating autonomous actors, eliciting spontaneous opinions and making their authentic feelings calculable and comparable with those of others. Resistance against such techniques of governing can take equally paradoxical forms, such as a refusal to be spontaneous and authentic in a dialogue with the human resources person or the desire to be helpful in a focus group or the wish to be a 'good subject' in a psychological experiment. Because of these paradoxes, tact is an all-the-more-important part of techniques of government. Tact is required to contend not only with the vagaries of fortune and the diversity of people but also with the tensions inherent in such systems of power.

Calculation

At this point it is worthwhile to return to Billig's argument, discussed above, that rhetoric can never be formalised into a finite system of rules based on psychological laws because there may always be something novel in a rhetorical situation. Quintilian's uncertainty principle, as Billig calls it, means that rhetoric always requires judgement and can never be a matter of mechanically applying the rules that pertain to a particular constellation of speaker, audience, message and context. Rhetoric, one could say, is not fully calculable. The relation between tact and technology in modern forms of governing, however, also shows that the tactful, kairotic aspects of governing are intimately related to the calculative, technological aspects. Tact and technology are not simply different, mutually exclusive ways of doing. Tact in the examples mentioned above (the focus group and management by dialogue) functions as a social lubricant for systems of calculated management and marketing. Its role is in part to mask the ulterior goal of the process, to hide the cool calculation behind a human face. Moreover, although tact is not fully calculable, there are guidelines and principles to point the manager or moderator in the general direction. Tact is not simply external to calculation.

Sociologists Michel Callon and John Law, discussing the relation between calculation and non-calculation, have argued that the boundary between them is always complex.[79] They start off by making the point that the important difference is not that between calculation and judgement. What happens in a court of law, for example, may not depend on the manipulation of numbers, but it nonetheless shares structural characteristics with quantitative calculation.[80] In both cases, relevant elements (measurements or evidence, for example) are sorted and displayed together in a single space, relations are created between them, and a conclusion is drawn (a result, a judgement). Reaching a legal judgement can be as orderly a process as arithmetical calculation.

Once one accepts judgement as a (qualitative) form of calculation, Callon and Law argue, the most important boundary is the one between arrangements that allow calculation and those that prevent it. Many religious practices are examples of the latter. In Quaker worship, for instance, the goal is to make oneself receptive to the Holy Spirit, which involves an active letting go, letting things happen without passing judgement. To reach such a state of passion is work, the active resistance of calculability. How to do so is spelled out in Quaker texts as a number of principles and procedures. During worship, for example, people are mostly silent, but if one is 'moved to speak', one may. The others are supposed to reflect on your words, but debate is strictly forbidden. Resisting calculation in this case is a matter of 'calculative rarefaction', of arranging a space in which passion, non-calculation, may occur.[81]

Callon and Law emphasise that there is no general rule for the relation between calculability and non-calculability. The boundary between the two is always a space of simultaneous connection and demarcation, and the proportions may vary. The gift is a good example: gift-giving is at 'a nexus of tensions between two regimes', namely that of pure brotherly love and that of economical exchanges.[82] Depending on how it is enacted, gift-giving will be more or less calculated – more like a market transaction or like an act of love – but it always sits between and combines aspects of two regimes that are necessarily in tension. Similarly, tact can be more or less strategic (or perhaps 'tactical' would be a better word here). Tact is always situated in the paradoxical space between calculated manipulation on one side and spontaneity on the other. Where it sits in that space is an empirical matter that I explore in this book.

Human Engineering

Finally, a word about the term 'human engineering'. It is tempting to classify techniques such as those of modern government as a separate kind of technology. Rose, for example, has distinguished human technologies from material technologies. 'Human technologies' are deployed by 'engineers of the human soul'[83] as instruments for 'linking together, shaping, channelling, and utilising the forces of individuals and groups in pursuit of certain objectives'.[84] They are hybrid technologies, assemblages of heterogeneous elements including architecture (a school building, for example), instruments (such as an intelligence test) and people (teachers).[85] What sets them apart from material technologies is that they aim to transform human conduct. Similarly, sociologists of science Trevor Pinch, Malcolm Ashmore and Michael Mulkay noted that technology studies remain dominated by a rather conventional view of technology as a matter of machines, artefacts and devices.[86] To correct for this bias, they proposed the term 'social technology' for technology that 'has its origins in the social sciences, and that although it may incorporate some material artifacts such as computers, ultimately its purpose is to produce changes in human behavior'.[87]

This way of distinguishing 'human' or 'social' technology – as technology that is aimed at modifying human behaviour – is problematic, however. Firstly, as scholars of science and technology studies (STS) have shown, devices are often used in a different way than they were intended to and may have effects that were never foreseen by their designers. What a device ends up doing is the result of the interaction of many actors, including users as well as designers and producers. Thus, if we distinguish a kind of technology by the aims behind it we ignore most of its existence, which may be very different than was planned. Secondly, all technology has 'social' components and effects, a point that is sometimes abbreviated as 'all technology is social'. As we noted above, this has been one of the most fruitful aspects of the approach to technology in recent STS.

In addition, terms like 'human engineering', 'social technology' and 'social engineering' are themselves part of the history I am describing. Throughout the twentieth century various social scientists have put great store on distinguishing 'human' or 'social' forms of technology from material or physical technology. It was an important vehicle for promoting their expertise: by suggesting it was as powerful as physical machines and/or by claiming it was a necessary complement to physical technology to offset its de-humanising effects, for example. Other scholars rejected human technology as itself an inhuman, degrading approach to human behaviour and social affairs. To these people, technology was at odds with what is essentially human, a threat to human values, and human technology was therefore doubly monstrous. One way or another, to both proponents and critics, the distinction between physical and social technology (and similar distinctions) has been highly significant, a site of contention over essential aspects of modern life. The history of such distinctions is a process in which what it means to be human, what a society is and what technology is, is variously articulated.[88]

In other words: concepts like 'human technology' and distinctions between human and physical forms of engineering are an important part of the history of what I have called the dream of efficient control over human behaviour. Above, I have referred to that dream as 'human engineering', but such terms, although useful as a shorthand or placeholder, are themselves part of the history that I explore. Rather than to adopt or reject them, I study them as such and explore the hope expressed in them and the resistance they have encountered.

3 Scientific Management and the Human Factor

> We are not mere machines; we are human beings, and protest against being discussed and considered as coequal with machinery.
>
> —James Duncan[1]

> The human factor cannot be ignored in the industries.
>
> —John Frey[2]

In the nineteenth century, the dream of control over people increasingly became couched in terms of science. Already in the 1780s French scholars had begun to discuss the possibility of applying the methods of natural philosophy to the study of social affairs. The term 'social science' appears to have its origin in the intellectual ferment of the French Revolution.[3] From the start, the betterment of society along rational lines was the main aim. Condorcet, for example, proposed that justice, prosperity, happiness and equality between individuals and nations could be attained by a 'social art', in a way that was entirely analogous to the improvement of human kind's material conditions by applying knowledge of the laws governing the natural world. Mastering the complexity of social affairs by reduction and classification and the 'application of the sciences of calculation' would allow scientific progress in the study of society, and the judicious application of this knowledge would result in 'the absolute perfection of the human species'.[4]

The idea that a better society could and should be constructed by applying a science that was modelled on the sciences of the physical world survived the French Revolution (unlike Condorcet himself) and entrenched itself in the social sciences that developed in the nineteenth century. According to Latour, social engineering 'has formed the core vocation of most social sciences' ever since.[5] A fundamental tenet of this ambition is that the structure of society is only truly visible to 'the keen eyes of the trained social scientist', not to the people themselves, and that a better, more rational society can therefore only be constructed on a scientific basis, not through political means.[6] Peace and prosperity, if not the absolute perfection of the human species, depend on recognising the essential ignorance of ordinary citizens and letting science decide what is best for them and how to attain it. In the next chapter I will describe several variations on this ideal as they were pursued by sociologists in the first half of the twentieth century and the role that the concepts of 'social technology' and 'social engineering' played in their discourses. In this chapter I focus on the introduction of an engineering approach to management by the American engineer and management consultant Frederick W. Taylor.

Shop Management

In Taylor's 'scientific management' the relation between the expertise of the manager and the ignorance of the worker had a dynamic quality: it was, to a large extent, purposely constructed by the management and actively resisted by the workers. In Taylor's calculations, the actions of the workers were not essentially different from the movements of the machines, and likewise their control was to be a task of the management, not of the workers themselves. The controversy over this approach to shop management spurred the articulation of what was sometimes called 'the human factor'. Taylor's inclusion of workers in the ambit of his scientific management – the exact way they performed their job, how they worked together and with their machines, how, when and how much they were paid – raised the question of whether one could treat people on a par with the machines in the shop. To bolster their position, proponents as well as critics of scientific management formulated ideas about the human factor, and in particular about what makes people different from or similar to machines.

Although the debate was at times almost philosophical, the point was always practical. How should men and machines work together? Are workers becoming the 'living appendages' of machines, as Marx had written? And how should the work in factories of ever-increasing size be managed? What sort of practice should that management be? Is it a kind of engineering? If not, what is it then? At issue, in other words, were the relations between workers and machines as well as the relations between the various people in the shop (owner, managers, foremen, workers) and how these relations should be governed.

Psychologists, starting with Walter Dill Scott and Hugo Münsterberg, succeeded in positioning themselves as the experts of the human factor and their instruments and expertise as the necessary complement of Taylor's psychologically flawed system. Traces of this boundary work can still be found. For example, Thomas Hughes sides with Taylor's critics when he writes that 'Taylor tried to systematize workers as if they were components of machines', judging him to have been 'naïve in his judgments about complex human values and motives'.[7] Ludy Benjamin and David Baker similarly state that Taylor ignored 'human issues', issues that were 'the domain of psychology'.[8] John Benjafield compares Taylor and Elton Mayo, writing that 'Taylor's focus was on individual behavior seen as a collection of individual bodily movements, while Mayo's focus was on behaviour as determined by the quality of one's interpersonal relationships'.[9]

Such contrasts ignore on the one hand the human factor in Taylor's thinking and early industrial psychologists' mechanistic view of the worker on the other. In some ways, Taylor's ideas about the worker's mind and about the right way for managers and workers to work together were similar to the 'human relations' approach of which Elton Mayo was a major proponent. This approach started making its name in the 1930s, long after Taylor had died. For their part, early industrial psychologists presented themselves as human engineers, offering techniques to measure and manipulate the mind of the worker. These developments are better understood if we do not take the 'human factor' as a given, ignored by Taylor but gradually uncovered by psychologists. The history of management from Taylor to human relations is not one of progressive humanisation but rather one of successive redefinition and realignment of the human and mechanical factors of production and their management.

The way in which the human factor was conceptualised, in particular how it was similar to or different from machines, developed in close connection with ideas about its management, focusing on the question of whether management was a form of engineering. In my interpretation of this history, I revisit well-known terrain. I draw mainly on Taylor's work and that of Lillian Moller Gilbreth and Mary Parker Follett and on the hearings regarding scientific management of the US House of Representatives. The main focus of this chapter is the period between 1910 and 1920, during which the relation between men and machines dominated the discussion about labour and management.

The 'Labor Problem'

When Taylor wrote his first essay on management, 'A Piece-Rate System', he gave it the subtitle 'Being a Step toward Partial Solution of the Labor Problem'.[10] American industry was rife with friction in the nineteenth century, and in the 1870s the term 'the labor problem' began to be used to refer to the tensions between capital and labour, and the unrest – strikes, boycotts, violence – that they gave rise to.[11] Social relations on the shop floor were an important aspect of the labour problem. The growing scale of industry and the increasing mechanisation had separated the factory owners from the production process. On the shop floor, the foremen were the dominant figures, often de facto subcontractors. Many foremen abused their power by engaging in favouritism and 'driving', spurring on workers to a high pace with a combination of authoritarianism and physical force.[12] Large-scale industrial unrest often arose out of conflicts between foremen and workers.

One approach to the solution of the labour problem was labour reform that focused on improving the relations between employers and workers and the working conditions of the latter. Such 'welfare management' could include measures ranging from the provision of a clean and safe working environment to sports clubs and cultural events and pension and insurance plans. Taylor's 'Piece-Rate System' fell into another category of approach to the labour problem, one defended in particular by engineers, who targeted the restriction of output that they saw as the most pressing part of the labour problem. Taylor's essay, a paper he had presented to the American Society of Mechanical Engineers (ASME), proposed an improvement on the wage-setting schemes that had been devised by Henry Towne and Frederick Halsey. Each of these three engineers based his ideas on the assumption that workers are primarily motivated by the desire to increase their wage, but their schemes were often rejected by workers, who still 'viewed the union as the proper antidote to the foreman's abuses'.[13]

Frederick Winslow Taylor (1856–1915) came from an upper-middle-class Philadelphia family. Although he passed Harvard's admissions examination, he had his eyes on engineering and became an apprentice pattern maker at a foundry – getting his hands dirty with the ordinary workmen, as he would later often remind his critics. In 1878 he moved to Midvale Steel in Philadelphia, where he soon became a 'gang boss', the start of a career in management. At Midvale, Taylor distinguished himself with his experimental attitude, leaving no element of the production process unexamined in search of improvements. Apart from such aspects as the proper tools and machine speeds and the routing of materials and products through the shop, payment schemes and their influence on worker productivity had Taylor's particular attention.

When he presented his 'Piece-Rate System' to the ASME in June 1895, Taylor had become a management consultant, and the enthusiastic reception of his paper brought new clients to him. In 1898 Taylor's fame and that of his management system got him a contract with the Bethlehem Steel Company in Philadelphia, his most important client yet. At Bethlehem, Taylor and his assistants had to grapple with the resistance of both managers and workers to the changes that Taylor tried to introduce. Nevertheless, when he left in 1901, his authority fatally undermined by the constant obstruction, he could claim a number of successes. Moreover, high-speed steel, an invention he had made at Bethlehem with his assistant Maunsell White, made them a fortune, and Taylor quit the consultancy business to become a full-time propagandist for his own management system, leaving the consultancy to his disciples.

Taylorism

The core of Taylor's shop management could be said to consist of four elements: a focus on individual workmen; a strict separation of the planning and the execution of work; the exact timing of the constituent parts, the elementary units, of a job; and finally his differential piece rate. The first two prepared the ground for the last two. According to Taylor, managing labour scientifically required individualising it and concentrating its management in a dedicated department, workers generally being incapable of understanding the science behind their own work. With the work itself rendered calculable by analysing it into unit tasks performed by individual workers, one proceeded to measure the time it takes a 'first class man' to do a particular subtask (such as getting one shovelful of coal into the boiler), if necessary improving the efficiency of work by removing wasteful, unnecessary motions. By adding up the unit times for each job, Taylor then determined what the ideal performance of his workers was. Workers who performed at this level were guaranteed a piece wage set 30 to 100 per cent higher than the average in the trade – those who failed were paid a wage somewhat below average. Because the resultant increase in productivity was higher than the increase in labour costs, the result was, in modern terms, a win-win situation – according to Taylor and his followers at least: higher wages and lower relative labour costs.

Through Taylor's proselytising and the consultancy work of associates like H. L. Gantt and C. Barth, the fame of 'Taylorism' continued to grow in the first decade of the twentieth century. Although Taylor did not realise it himself at first, his ideas about management, and in particular the role of technical experts in organising the processes in a factory, resembled the ideas of Progressive thinkers about the reform of society.[14] The big breakthrough came in 1911, when the progressivist lawyer Louis Brandeis, representing before the Interstate Commerce Commission a group of trade associations that opposed the rate increases that the railroads were threatening to introduce, made 'efficiency' the central theme in his case. Almost overnight Taylor became a public figure and 'scientific management', a catchy name devised at Brandeis's request, became a household word. An efficiency craze that would last several years had begun.

Not everyone was enthusiastic about scientific management, however. The Eastern Rate Case, as it became known, also galvanised the hostility towards efficiency measures that many workers and their union representatives harboured. Bonus and other wage systems in particular had long been met with resistance because they were seen as forms of 'driving' or 'speeding': once workers had been tempted into higher productivity by some form of piece-rate or bonus system, employers would often cut the rate, and then workers would have to keep up the same high pace in order to make their old, low wage. The unions' 'abhorrence of rate-cutting was matched only by their distaste for the "speed-up"'.[15] For his part, Taylor was staunchly anti-union. Scientific management, he claimed, allowed management to set tasks, working hours and wages in an objective, scientific fashion, and removed the need for unions and collective bargaining. Management should be a form of engineering, not politics. With the spotlight on scientific management after the Eastern Rate Case, the unions mounted a counter-attack.

Workers and Machines

It was Samuel Gompers, president of the American Federation of Labor, who led the charge. In February 1911, with the ICC about to rule on the rate increases, Gompers wrote the first of several vitriolic pieces against scientific management. In his editorial in the *American Federationist* Gompers called it a 'machinery to perfect the living machine'.[16] Two months later, in a longer essay reacting to the first instalment of Taylor's 'Principles of Scientific Management' in *American Magazine*, he repeated the charge: Taylor wants to make 'high speed automatic machines' out of workers, and fit them into the shop as 'a cog or a nut or a pin in a big machine'.[17] Gompers was not the first to employ this trope – a few months earlier naval constructor H. A. Evans had warned his fellow machinists that the Taylor system 'is devised to make machines of them and cheap machines at that'.[18] Both Gompers and Evans were probably inspired by Marx's characterisation, in *Capital*, of the difference between manufacture and the factory. Whereas in manufacture the workers are part of a living organism, in the latter 'we have a lifeless mechanism which is independent of the workers, who are incorporated into it as its living appendage'.[19] Gompers had more clout with American workmen than Evans or even Marx (Gompers himself was not a Marxist), and it was his rhetoric that set the tone for the controversy about scientific management.

As the fame of Taylorism spread, the charge that it reduces workers to machines seemed to follow in its footsteps. Alphonse Merrheim of the French Fédération des Méteaux denounced Taylor because 'the worker must become, he has become, an automaton'.[20] E. Pouget of the French General Federation of Labour concurred: '[I]t makes the worker an automaton, destroys his initiative, renders superfluous any effort of thought or reflection, and in brief makes him an accurate and unthinking extension of the machine.'[21] The Dutch Jesuit J. van Ginneken, who in 1917 took it upon himself to save Dutch Catholics from the scourge of Taylorism, warned them that it treats the worker 'as a merely material machine, the sole purpose of which is to yield the highest production'.[22] The trope had two variants: sometimes it had the worker reduced to a machine, as in the examples above, and sometimes the original, Marxian version was used, and the worker had been made part of a machine, as when John Frey of the Molders Union claimed that 'scientific management as applied to workmen moves with the smoothness of a well-oiled and perfected machine, in which each one performs his part with the accuracy of a mechanically and mathematically perfect tooth on a gear wheel, when it meshes with the teeth of another wheel in transmitting power'.[23] The difference – machine or machine part – was of no consequence in the discussion.

The element that distinguished men from machines, that which scientific management deprived the workers of, was articulated in many different ways. Gompers claimed that separating planning and execution robs the men of 'initiative'[24] and that it is a system that 'preys upon the independence, the development, and the character of the worker',[25] his 'individuality, initiative and vitality'.[26] The AFL's vice president James Duncan wrote, in the *American Federationist*, that a task-setter would rob him 'of an individuality and self-reliance and reduce me to an automaton'.[27] Apparently taking his cue from Gompers, Merrheim rued the fact that 'the personality, the intelligence, even the desires of the worker are destroyed'.[28] Van Ginneken simply called it their 'soul'[29] but also pointed out that workers were stripped of their 'free will'.[30] Commenting in 1933 on Soviet Russia's appropriation of scientific management,[31] the philosopher and social activist Simone Weil still lamented the way it made living cogs out of skilled workers with freedom, intelligence and initiative.[32] All in all, it was a common criticism that Taylor's system stripped workers of their humanity and reduced them to machines, but what humanity consists in was formulated in a rather diffuse although familiar set of terms such as intelligence, personality, initiative, freedom, soul and individuality.

Scientific management, including the question of men and machines, soon became a federal matter. In April 1911 representatives of the machinists of the Rock Island arsenal came to Washington to protest against plans to introduce scientific management. Taylorism, they complained, is an attempt to speed up the worker to a 'terrific pace' by methods including '[s]tandardizing the movements of a workman, thus making an automaton of him'.[33] In response, the House of Representatives passed a resolution asking for an investigation. In the preliminary hearings, Representative Irvin Pepper, who had introduced the resolution, stated that '[t]he government should not approve a system which reduces the laboring man to a mere machine'.[34]

When later that year a strike broke out at the Watertown navy shipyard over the introduction of scientific management, the House set up a special committee to investigate 'the Taylor and other systems of shop management'.[35] The committee, made up of Representatives William B. Wilson, William C. Redfield and John Q. Tilson, heard a long list of critics and proponents of scientific management, including Taylor himself. The charge that Taylor's system made machines out of workers had their special interest. Chairman Wilson in particular was convinced that Taylor's system makes 'the workmen nothing more nor less than human machines to carry out the instructions that other intelligence has evolved'.[36]

The Mechanism and the Essence of Scientific Management

Clearly, Taylor and his associates needed to defend their work against the charge that they reduced workers to machines. Taylor developed two strategies. In his 'Principles of Scientific Management', written around the time of the Eastern Rate Case, Taylor introduced an analogy: was the surgeon, trained to work with precise motions and according to strict procedures, any less human?[37] When Wilson put the automaton criticism to him, Taylor used the same analogy: trainee surgeons have to do exactly as they are told, yet we don't consider them to be automatons. His associate Frank Gilbreth took the argument one step further. After publishing 'Principles of Scientific Management', the progressive journal *American Magazine* received hundreds of letters from fascinated but more often worried and indignant readers. Taylor delegated the task of answering them to Gilbreth, who wrote a *Primer of Scientific Management*.[38]

The *Primer* is written in question-and-answer form, and one of the questions is 'does it not make machines out of men?'[39] 'Is a highly trained sportsman or soldier a machine?' Gilbreth asked in return. He is simply trained to perfection. If it means doing the work in the one best way, then yes, scientific management tries 'to induce men to act as nearly like machines as possible'.[40] And such 'machines' (Gilbreth appears to mock the charge) are paid handsomely for their efficiency. '[W]hether or not the men may be called machines, they fare better and profit more.'[41]

However, in the hearings Taylor's main line of defense follows another strategy: emphasising a distinction between the 'mechanism'[42] of scientific management and its 'essence'.[43] Right from the start of his opening statement, Taylor takes great pains to convince the members that there is more to scientific management than mechanism and efficiency, that the precisely timed routine movements, the standardised tools, and the system of payment that it is famous for are only possible on the basis of a 'complete mental revolution'.[44] Taylor apologises for sounding a little 'highfalutin',[45] but the essence of scientific management lies in the 'atmosphere' surrounding it, the 'sentiments' accompanying it.[46]

Science is one aspect of this essence: scientific management does not consist in the specific techniques that are employed at any one time and still less in the techniques that Taylor himself happened to have devised but rather in the scientific attitude of which they are the result. If scientific experimentation proves that there is a better way to organise work, any technique will instantly be dropped in favour of it. Taylor has much more to say, however, about the other element of the essence of scientific management. This part of the mental revolution consists in workers, managers and employers substituting 'friendly cooperation and mutual helpfulness for antagonism and strife'.[47]

Then even more than now, there was ample reason to be sceptical about the idea of friendship between factory owners and the men and women who work for them. Taylor himself treated the theme to some extent as a ploy, describing it in a letter to another witness as 'the most effective weapon that we have all of us found as a foil against inconvenient questions'.[48] It was a theme, however, that went back a long way in Taylor's work before the hearings.[49] Reading *Shop Management*, in which Taylor first expanded his ideas about the differential piece wage into a comprehensive management system, one is struck by his empathic analysis of the social and psychological processes at work in industry. Ultimately, the reason why scientific management is necessary is the tremendous waste in industry, its inefficiency, and the root cause of that is soldiering, deliberately working at a slower pace than is possible.

Soldiering proceeds from two causes, Taylor explains: from the workers' natural tendency to take it easy, which they have in common with everyone else, but also from 'more intricate second thought and reasoning caused by their relations with other men, which may be called systematic soldiering'.[50] Workers realise it is not in their interest to work harder than another for the same pay, and they also know (and if they don't, more experienced colleagues will tell them) that employers will rarely reward higher production with more pay. If the work is done for a piece wage, the rate will be cut when production goes up. Thus, workers carefully determine the precise minimum pace of work that they can get away with and make sure their employer remains convinced it is the maximum pace by disciplining whoever exceeds it.

To prevent soldiering, the manager must appropriate the power to plan the job and employ time studies to determine the proper pace. However, the essence of scientific management, Taylor insisted, lies not in the use of the stopwatch but in the relationship between managers and workers, which should be one of cooperation, friendship and teaching. Strife has been the main organising principle in industry, according to Taylor. Employers want low labour costs, workers want high wages, and from the apparent incompatibility of these two interests the main traits of contemporary industrial life follow: labour unions and strikes, soldiering by the workers, wage cutting by the employers, driving by the foremen. This strife is inefficient, but it is also unnecessary. Low labour costs and high wages are not incompatible at all. If the work is done in the scientifically determined most efficient way, the employer can pay high wages to fewer workers and still raise production, and the worker can earn more without being driven to exhaustion. Once they realise this, employers and workers can shift their attention from dividing the surplus to increasing, hand in hand, the size of the surplus.

This principle of profit through peace and cooperation cannot exist by its rationality alone. It has to go deeper than mere calculation. There must be real friendship between employers, managers, foremen and workers to make scientific management work. '[S]mall acts of personal kindness and sympathy' work wonders to 'establish a bond of friendly feeling',[51] but companionship is sustained to an important degree by 'daily intimate shoulder to shoulder contact'.[52] Although scientific management pulls apart the thinking and doing of labour and has them embodied in different people, these people (managers and workers) should be together on the shop floor as much as possible. '[M]ore than all other causes, the close, intimate cooperation, the constant personal contact between the two sides, will tend to diminish friction and discontent.'[53] On the shop floor, managers and workers perform a precisely choreographed dance, the actions of the one dovetailing with those of the other. On the one hand, this close cooperation requires good relations, but on the other hand in such circumstances 'it becomes practically impossible to have a serious quarrel'.[54]

Essential to the relations between management and workers is honesty. In the old days in industry, deception and deviousness reigned in the battle between employers and workers. If workers knew there was a quicker way to do a job, they would never tell management, and employers could not be trusted to keep their promises and not cut piece rates at their own convenience. Taylor, on the other hand, tells the manager always to be 'perfectly straightforward' with the men.[55] Time and motion studies should preferably be done openly, by fair-minded men, with full consent of the workers. Once a differential piece rate is set, it must be religiously adhered to. In return, the manager may expect fairness from the workers. And indeed, as long as the management does not engage in 'sneaking business', the men will cooperate honestly.[56] Such is human nature.

To bring about the mental revolution on which scientific management should be based, the manager must be an orator and a teacher. Long passages in Taylor's works are devoted to descriptions of how a shop should be converted to scientific management. The technical, mechanical part of this process is a relatively minor concern. It is the mental part, winning over the minds of workers, owners and managers, that makes it such a difficult and protracted affair – one that may take years to complete. Both employers and workers must come to see the benefits of the new system despite the drawbacks it may at first sight appear to have. Convincing the employers is a matter of patient explanation, but the rhetoric for the workers is built primarily on 'object lessons'. One should start by converting one man to the new system. Some coercion may be required at this point: the worker is given the choice to cooperate or leave. Only when this man has learnt to work in the new way, and has realised that it pays to cooperate with management under its regime, can one introduce other workers to it one by one. Once about a fourth to a third of the workers have been brought over, the tide will rapidly turn, but it is imperative not to rush at first.

Although object lessons are the main rhetorical instrument, there is nevertheless a lot of talk involved in changing the minds of the workers. Taylor's preferred rhetorical form for this 'difficult and delicate task' is the one-to-one conversation, in which the characteristics and the advantages of the system are patiently explained and the worker has the chance to ask questions.[57] A lot comes down to the precise way the men are addressed. Taylor's favourite example was his conversation with 'Schmidt', a worker who was to be taught the scientific way of handling pig iron. In the talks Taylor held in his mansion for entrepreneurs, engineers and politicians, he would re-enact the scene complete with imitations of Schmidt's mental sluggishness and Pennsylvania Dutch accent ('Vell, I don't know vat you mean').[58]

More talk is required in the instruction that is an integral part of Taylorism. With planning and execution separated, it becomes essential to communicate the ideas of the planners to the workers so that they will execute the job in the manner and time that has been scientifically determined for it. Specially assigned 'functional foremen' teach each worker the proper way of doing a job, find out what's wrong when their production falls behind, encourage them and discipline them if necessary. Time and again Taylor emphasises that it is the responsibility of management not only to select the best worker for each task but also to train and develop them to do the work as required. 'It is only when we fully realize that our duty, as well as our opportunity, lies in systematically cooperating to train and to make this competent man, instead of in hunting for a man whom some one else has trained, that we shall be on the road to national efficiency.'[59]

To an important extent, Taylor's discourse about the essence and the atmosphere of scientific management was mere talk. Historian Daniel Nelson has pointed out that in practice Taylor was not much interested in personnel management, had a strong authoritarian streak in the way he dealt with his workers and preferred to consider them in terms of the amount of 'foot-pounds' of work they could be coaxed to do for a certain wage.[60] Selection and training were of little actual importance or took the form of weeding out the unfit and disciplining the slow and lazy. Nelson, however, has also shown that something similar is true of the mechanism of scientific management: that too remained to some extent mere theory. What was implemented of it in practice did not differ much from the 'systematic management' that was already widespread. The measures that were introduced under the flag of scientific management often did not include time study, for instance.

Taylorism was significant first and foremost as an idea. In this respect the 'mental revolution' was no different from time study or the differential piece wage. In any case, Taylor's defence of scientific management in his testimony at the hearings was not successful. Rep. Wilson was not impressed by the comparison of a scientifically managed worker with a surgeon. The point is not, he said, that trainee surgeons, like workers under scientific management, are taught the current best practices. If they were scientifically managed, surgeons would not even be allowed to perform a diagnosis or determine for themselves the best way to operate; they would be given an instruction card from the surgeon in chief and be expected to follow it to the letter. In its final report the committee underscored the distinction between people and machines: 'A machine is an inanimate thing – it has no life, no brain, no sentiment, and no place in the social order. With a workman it is different. He's a living, moving, sentient, social being . . . He would be less than a man if he did not resent the introduction of any system which deals with him in the same way as a beast of burden or an inanimate machine.'[61] Moreover, the committee gave Taylor's fancy idea of a mental revolution short shrift, arguing that a spirit of collaboration is 'too variable and insubstantial a basis upon which to rest the material welfare of the wage worker'.[62]

Lillian Moller Gilbreth's Psychology of the Human Factor

At the time the controversy over scientific management broke out, the discourse about industry already contained a term that indicated the role of labour in production as opposed to machinery: the 'human factor' or 'human element' (the two terms were synonymous).[63] If one needed to point out, for example, that industry depends on workers, not just on machines, or that human labour is more important than machinery, one could refer to the 'human factor'. Although he didn't use the term 'human factor' himself, Gompers's charge that Taylor made machines out of men dovetailed with a distinction that was already in place. Others explicitly articulated Taylor's faults in term of the human factor: Taylor overlooked the human element because he didn't appreciate the sensibilities and expertise of the workers,[64] because he preferred automatons to skilled men[65] and because he considered workers as merely mechanical parts of the machinery of production.[66] The human factor came to stand for the wide array of aspects that distinguished labour from machinery, men from machines, and that was ill served or even effaced by scientific management.

Scientific management's perceived shortcomings regarding the human factor offered an opportunity for psychologists looking to develop psychology as a practical science. To increase efficiency in industry, it is necessary to consider not just the 'mechanical side of production' but also human efficiency, stated Walter Dill Scott, and this requires psychological study of the human factor.[67] Scientific management is useful when dealing with routine work, but it 'has done little for men whose welfare depends upon judgment' because they work in positions of responsibility.[68] Two years later, Hugo Münsterberg wrote that he believed the principles of scientific management would 'prove of lasting value'[69] but that he deplored its 'helpless psychological dilettantism'.[70] 'The theorists of scientific management seem to think that the most subtle methods are indispensable for physical measurements, but for psychological inquiry nothing but a kind of intuition is necessary.'[71] Others followed his example. Whether they called their field psychotechnics or industrial psychology, when they first entered the domain of industry psychologists articulated their approach and skills in relation to scientific management, though not all as negatively as Münsterberg.

In Lillian Moller Gilbreth psychology had a representative who identified with scientific management. She believed that psychology and scientific management were quite compatible, that 'Scientific Management as laid down by Mr. Taylor conforms absolutely with psychology'.[72] With her husband Frank Gilbreth, Lillian Moller (1878–1972) formed an efficiency team that was a considerable force in scientific management. In their work together Moller Gilbreth complemented her husband's energetic, practical attitude with solid methodological and theoretical expertise and did much if not most of the writing. Their work focused on the movements of the worker rather than on the pace of work. Applying motion studies to bricklaying had been a longstanding interest of Frank, and *Bricklaying System*, written by Lillian but credited to Frank, appeared in 1909.[73] By that time they had met Taylor and had been taken up into his circle of associates. They would remain on the periphery however. Taylor was ambivalent about Frank, didn't entirely trust him but considered him useful because of his considerable rhetorical talent. Moreover, the Gilbreths' efficient system of bricklaying became a staple of Taylor's own propagandistic efforts. Frank considered Taylor 'a very great man' but didn't yield to his authority.[74] To Taylor's increasing chagrin, the Gilbreths presented motion studies as an improvement on time studies. Lillian for her part never liked Taylor and had her doubts about the scientific character of his ideas. In 1914 Taylor had had enough of their independent attitude and excommunicated them from the movement. The Gilbreths, who in 1912 had started a scientific management consultancy firm, were on their own and began to cultivate an image of 'the good exception': scientific management but with an eye for 'the human factor'.[75]

Lillian's work on the psychology of management was an essential part of the Gilbreths' specific brand of scientific management. After getting to know scientific management through Frank, she had chosen to do a doctorate in educational psychology, reasoning that it offered a way into professional management and would give her expertise in an aspect that people like Taylor tended to neglect.[76] Taylor in fact was so impressed by her work on the psychology of management that he mentioned it in his keynote speech to the 1910 conference of the ASME and the British Institution of Mechanical Engineers.[77] Berkeley, however, denied her the PhD degree because she hadn't spent the last year of her doctoral programme on campus. She decided to publish the book and look for another institution.[78] At the request of its publisher, Moller Gilbreth published *The Psychology of Management* as 'L. M. Gilbreth' so as not to hurt sales by divulging it was written by a woman.[79] There is a sad irony in this, as she would arguably become more important in the history of psychology and management than Frank Gilbreth but also in the light of the importance she assigns in her book to individuality.

In frequently extravagant terms Moller Gilbreth impresses on the reader the importance of recognising and acknowledging each worker's unique personality, making sure that the work he does is adapted to it and arranging the processes in the shop in such a way that the personalities of the workers are allowed and encouraged to develop to their fullest. Unlike Münsterberg and most other psychologists of her day, Moller Gilbreth felt no need to contrast scientific management and psychology; in fact '[t]he psychological element of Scientific Management is the most important element',[80] as it is built on the recognition of the individual not just as an 'economic unit' but also as a personality.[81] Moller Gilbreth made the core of scientific management psychological.

Moller Gilbreth repeatedly addresses the charge that scientific management reduces workers to machines. This is not so, she counters; on the contrary, it encourages and helps the worker's personality to develop. It starts with careful, scientific selection, which assures that each worker is perfectly suited for the work he does. '[F]ar from being "made machines of", men are selected to reach that special place where their individuality can be recognized and rewarded to the greatest extent.'[82] Once in that special place, each worker should regularly receive a record of his output, preferably relative to others, encouraging a feeling of personal responsibility, fostering competition and impressing upon him that he is an individual, not an anonymous member of a gang. Again, '[t]his chance to be an individual, or personality, is in great contradistinction to the popular opinion of Scientific Management, which thinks it turns men into machines'.[83] The effect is even greater if the worker makes his own records, which requires him to be 'conscious every moment of the time exactly where he stands' and thus creates self-knowledge.[84] Somewhat counter-intuitively, the standardisation of work processes that is so typical of scientific management also supports and strengthens individuality. It conserves the mental energy of the worker and allows him to focus on the parts of the job that are novel. He can then use his initiative to come up with a good way of standardising them.[85] In fact, Moller Gilbreth continues, since standards are a way of externalising memory, an ability that distinguishes us from the lower animals, those who object to standardisation apparently want to turn the worker into an animal. Against all appearances, '[s]tandards prevent men from becoming machines'.[86]

Moller Gilbreth thus translated the human element, that which distinguishes us from machines, into a psychological factor. What had been a diffuse set of terms in the criticism levelled at Taylor – brain and intelligence, initiative, free will, personality, soul – she articulated as one factor, individuality, which could be scientifically measured and managed with the help of psychology. Scientific management that is based on psychological laws and principles selects the individual best suited for each task and organises the shop in such a way that each individual is enabled and encouraged to develop to their fullest. In this way, Moller Gilbreth put forward psychology as a mediator between the mechanical and the human: by employing psychology's knowledge of the laws that govern 'the working of the mind' one enables the worker to develop as an individual and be unlike a machine.[87] Psychology connects the domain of causal laws with the world of feelings, morals and imagination, feeding the latter with its knowledge of the former.

Although it is individuality that makes us different from machines, there is a second theme in the book that has less to do with individuality than with the relations between people. Moller Gilbreth adopts Taylor's emphasis on teaching and gives it a psychological basis. Teaching is ubiquitous, in industry as well as in the academic world, and it must be made the object of scientific management. '[E]very man going out into the world needs all the knowledge that he can get as to the working of the human mind in order not only to give but to receive information with the least waste and expenditure of energy.'[88] To make teaching efficient, it must be based on psychological principles. The 'disciplinarian', one of eight kinds of foremen that Taylor had distinguished, must 'apply, at least unconsciously, and preferably consciously, the known laws of psychology, if he wishes to be successful'.[89]

The laws and principles in question have a familiar ring to them: it is important, for example, to motivate the worker by explaining why he needs to know what he is taught, and it is crucial to gain his commitment by making him feel that he is 'a part of what is taught', that he is being taught the laws underlying his own experiences.[90] Moller Gilbreth combines such motivational advice with an emphasis on teaching the right habits, the psychological importance of which she supports by referring to William James. As in James's famous chapter of the *Principles of Psychology*, habit connects the structure of the brain with morality and a good and just society. On the one hand, the formation of proper work habits always starts with teaching the right motions (rather than abstract theory), resulting among other things in a fully developed brain, but on the other hand the benefit also lies on a higher plane, all the way up to the education of reason and will, the happiness of the worker and a Taylorist 'spirit of coöperation'.[91]

Engineering and Management

Between 1914 and 1920, the furore over scientific management gradually died down. Union resistance waned when more workers actually experienced scientific management and found out that time studies, touted by Taylor as the central part of the system, were in practice a minor element, most of the attention going to better management of machinery, routing, tools, shop layout, etcetera. As Daniel Nelson has discovered, this had been true of scientifically managed shops all along.[92] The House report of 1912 recommended that time study should only be done with the consent of the workers but did endorse standardisation, systematisation and other efforts to eliminate waste in industry and raise efficiency. Scientific management, minus time studies, was reintroduced in the US arsenals.

Meanwhile, with the passionately anti-union Taylor having died in 1915, his disciples sought closer links with union leaders. Taylor's pupil Morris Cooke edited a collection of essays together with Samuel Gompers and Fred Miller, the president of the American Society of Mechanical Engineers, in which 'the scientists of industry and the representatives of organized workers' together considered the problems of industry.[93] In the process, the 'workers are human beings, not machines' issue largely disappeared from the discussion of scientific management, instead becoming part of the common sense *topoi* of work on which Chaplin, for example, drew for *Modern Times*. At the same time, the dominance of engineers in management was gradually broken when psychologists and accountants began to compete for the same ground.

It would be tempting to interpret this process of gradually diminishing influence of engineers and a growing role of psychologists as the humanisation of management, culminating in the 'human relations' approach of the 1930s. The more psychologists had a say in management, it appears, the better the 'human factor' was represented, the more 'humane' management became and the less like engineering.[94] One reason why such a story would be incorrect is the psychological nature of some of Taylor's ideas. He never used the term 'psychology', but the mentality – the motives, feelings and thinking – of workers were of great interest to him. To say that Taylor was psychologically naive and Münsterberg and the psychologists who came after him increasingly sophisticated is to uncritically adopt the boundary work instituted by that same Münsterberg.

If anything, Taylor's perspective on the mind of the worker was broader than that of many early industrial psychologists and psychotechnicians. Whereas they focused on the individual worker – his or her personal traits and capabilities, their exact measurement using psychological tests – Taylor was equally interested in the social processes on the shop floor and in the relations of workers and managers. Some of Taylor's ideas have a distinct 'human relations' character to them.[95] In one of his favourite anecdotes, Taylor relates how his associate Sanford Thompson had reorganised the quality control in a factory producing ball bearings. The process had had a beneficial effect on the well-being of the 'girls' doing the inspecting. Not only could the work be done better, by fewer girls working shorter hours, it also raised their spirits: 'Each girl was made to feel that she was the object of especial care and interest on the part of the management, and that if anything went wrong with her she could always have a helper and teacher in the management to lean upon.'[96] Moreover, when the Hawthorne study revealed that workers restricted their output to a very precise level to maximise earnings while avoiding rate cutting, it replicated one of Taylor's basic insights into the causes of soldiering. The analysis of the group processes that underlie the phenomenon extended Taylor's but did not contradict it.[97]

Rather than management becoming more psychological with the advent of psychologists like Moller Gilbreth and Münsterberg, it is more to the point to say that the mind of the worker became technologically constituted by the theories and instruments they introduced: it became a machine working according to definite laws, a machine that could be manipulated using standardised techniques and the operating parameters of which could be determined using scientific instruments. Taylor already made some steps in this direction. In the *Principles*, he wrote that management should include not only the scientific study of work (task time, tools, etc.) but also 'the accurate study of the motives which influence men' because there too exist laws, even though human beings are very complex.[98] Psychologists took it upon themselves to formulate the laws that governed the mind and behaviour of the individual worker. What Moller Gilbreth writes about mind and management is not all that different from Taylor, except that she supports each claim with a reference to a psychologist's theory.

In one important respect, however, Moller Gilbreth's approach *is* different from Taylor's. Whereas Taylor had distinguished the mechanism of scientific management from its 'essence', the system from the atmosphere of cooperation and camaraderie that formed its heart as well as its precondition, Moller Gilbreth made the relational, emotional and moral aspects of management an integral part of the system and grounded them in the firm reality of psychological laws. She applied the precept of standardisation, for example, not only to the actions of individual workers but also to 'the relationships between the foreman, the manager, and the worker'.[99] As Moller Gilbreth's biographer Jane Lancaster puts it: 'Lillian, like all the scientific management experts, wanted a predictable and controllable labor force and was prepared to use her psychological expertise to get it.'[100] If anything, management was more like engineering in the theories of Scott, Münsterberg and Moller Gilbreth than it was in Taylor's work.

Around the Great War, engineering was a common term for the work of psychologists.[101] Applied psychology stands to experimental psychology as engineering to physics, Münsterberg proclaimed, and many expressed the analogy in terms like 'human engineering' and 'mental engineering'.[102] In an era when the engineer epitomised the ability to construct a better society for all, such terms conveyed the sense of professionalism that applied psychology required. '[P]sychologists capitalized on the hard-won reputation of the engineer as a practical man of affairs.'[103] Thus, with more psychology, management did not immediately become less like engineering. It widened the field of application of what was still seen as a technical practice. Complementing management with a concern for the 'human factor' did not make it more 'humane' as long as the human factor was still a kind of machine.

It is true that the mechanical model of the human factor was increasingly challenged in the interbellum and that psychology became an important source of resistance against mechanicism. The British industrial psychologist C. S. Myers, for example, who had been enthusiastic about scientific management in 1918,[104] soon began to protest against its mechanistic approach.[105] During the decade that followed, psychology channelled approaches into management that were incompatible with an engineering model, including instinct psychology[106] and Gestalt psychology.[107] Psychologists began to accuse scientific management of ignoring the complexity and uniqueness of people and organisations and professed the impossibility of analysing work and workers into their constituent parts. Thus, the increasing influence of psychologists on management thought and practice certainly developed the human factor into a much more diverse and complex aspect of work and management than it had been in Taylor's days, but the move away from mechanicism and engineering in management was the result of a change *in* psychology rather than being caused by the entry of psychology into management per se.

Managing Freedom and Responsibility

The increasing influence of psychology on management theory and practice continued a development that Taylor had started: psychologists extended the reach of science on the shop floor to include the individuality and the social relations of the workers. The influx of psychologists into management was consequential because it brought the human factor in industry under the aegis of science, taking management further in a direction that Taylor had already vaguely indicated. An arguably more important turn happened when the limits of the reach of science and engineering in industry were articulated. This happened when freedom and autonomy, rather than individuality, became the hallmarks of the human factor.

Of the diffuse set of terms that had been employed to mark the human factor, freedom and autonomy had been used to differentiate men from machines but even more to distinguish them from slaves. Didn't the worker under scientific management, forced to follow the instructions on an instruction card to the letter, his every movement precisely prescribed, lose his freedom and become a slave? The stopwatch, the object of much hate of the moulders and machinists in the hearings, was compared to the slave driver's whip.[108] Taylor strenuously denied the charge: the foreman in scientific management is not a 'nigger driver', he is a teacher, a friend even, there to help the worker do the job in the best possible way, earning him the most money.[109] Taylor emphasised that the system is sovereign in scientific management. If it had been objectively determined that the job had to be carried out in such a way in so much time, then everyone, including foremen and management, had to obey. Yet to workmen it was clear that science was a tool in the hands of management and employer, not vice versa. In practice, the boss remained sovereign, not science.

After the war there was a growing consensus that the old distribution of power in industry was unacceptable to workers and, moreover, did not 'work'. In the essays that Cooke, Gompers and Miller collected in 1920, the keywords that describe 'the human factor as the heart of industry'[110] are freedom, cooperation, responsibility and democracy, and the authors emphasise that only by respecting workers' freedom and encouraging them to assume responsibility for their work can productivity really be raised.[111] As 'labor manager' A. J. Todd put it, to really motivate workers to produce more they need to be able to assume 'industrial citizenship', and this means 'responsibility and some measure of self-determination and self-expression'.[112] In the words of Gompers: 'Liberation is the answer, and only through liberation can there be a solution of our problems.'[113] An early example of the style of management that such a conception of the human factor requires is described in Mary Parker Follett's 1925 paper 'The Giving of Orders'.[114]

Follett's interest in the social processes in organisations arose from her study in political science (she graduated from Radcliffe in 1898) and her experiences in social work in Boston.[115] She started to express her ideas in lectures and in 1918 published a book, *The New State*, in which she presented self-government arising out of group processes as the solution to problems of government. Shifting her focus to business administration, she began to reflect on the role of managers in cultivating, rather than imposing, change in organisations. In her ideas on this subject the psychology of human motivation and interaction becomes an increasingly important subject. 'The Giving of Orders', from 1925, is one of her best-known lectures on the topic. At a time when instinct psychology was under fire from behaviourists, Follett notes that the exaggerated interest in human instincts is giving way to a study of habits. Like Moller Gilbreth, she advocates creating habitual behaviour patterns in workers. With such 'prepared-in-advance behavior patterns' one prepares the way for 'the reception of orders' that will release these behaviours.[116] The subsequent reinforcement of the response completes the behaviourist scheme.

However, Follett warns, preparing in advance is not what psychologists take it to be, a tactic. People do not like to be told what to do; they have a deep-seated wish to be self-governing. This instinct, in fact, is 'the very essence of human being'.[117] A manager could only dispense orders in the way that psychologists envisage if it were a covert practice; otherwise employees would resist. 'If this prepared-in-advance idea were all that the psychologists think it, it would have to be printed privately as secret doctrine.'[118] Follett's way to avoid the Scylla and Charybdis of 'bossism' and avoiding orders altogether is to depersonalise the order.[119] Rather than the manager giving an order, inevitably eliciting resistance, the order is located in the demands of the situation, and the manager must engage the employee to discover it together. By putting everyone, manager as well as worker, 'under the situation' the problem of resistance to authority is overcome.[120] If it is the situation rather than the boss that rules, one may expect employees to follow orders and take responsibility for their work as well. Obedience and liberty come together under the 'law of the situation'.[121] Thus, Follett took Taylor's insistence on the sovereignty of the system and developed it further into the law of the situation. Not in the interest of science, as in Taylor's case, but to accommodate the essential autonomy of workers.[122] Control must give way to self-control, distributed over the whole company. 'To get our affairs in hand, to feel a grip on them, to become free, we must learn, and practice, I am sure, the methods of collective control.'[123]

Engineering is not a term that fits well such a form of management as Follett proposes. In their calculations engineers certainly need to factor in resistance, but the friction they deal with does not arise from the fact that their machines are both expected to follow orders and supposed to be autonomous. That, at least, was not the state of the art in mechanical engineering in the 1920s. Locating orders in the situation is a paradoxical instrument to manage the paradoxical challenge of governing autonomous subjects. In comparison, scientific management has a very straightforward idea of control. Defending scientific management in his *Primer*, Frank Gilbreth argued that it may be so that under scientific management you have to do the work as the planners have determined, but at least you get a 'square deal', and not being lied to is also a kind of freedom.[124] Although scientific management was in practice not free of deception (Taylor advised that time study was best done covertly if resistance was expected, for example), it is clear that this is a different idea of freedom than that of Follett. For Taylor and his disciples, freedom, initiative and independent thought had a place outside of the mechanism of objectively timed, precisely coordinated tasks. Inside the system one had to do as the system required. It was a transparent, honest, 'square' deal. In Follett's management, on the other hand, freedom and obedience are brought together in an uneasy combination. In the process, the subject under management changes in step: from a being that shuttles between machinality and friendship to a hybrid of freedom and responsibility, autonomy and order.

Tact and Engineering

Frederick Taylor is commonly portrayed as both psychologically naive and callous in his neglect of the human factor and in the way he treated workers as mere machines. Management only became humane with the arrival of psychologists on the scene. I have argued that this is too simplistic a view and uncritically repeats the rhetoric that early industrial psychologists like Münsterberg employed to position their profession. It assumes that the human factor is a given that was ignored by Taylor and uncovered and defended by psychologists. However, to understand this part of the history of management one needs to look at the ways in which the human factor was variously articulated.

Certainly, the core of Taylor's scientific management, its 'mechanism' as he called it, was a largely mechanistic affair, in which workers functioned primarily as bodies propelled by rudimentary motives, performing repetitive, precisely engineered tasks. But Taylor also emphasised that around this mechanism there had to be an atmosphere of friendship, honesty and cooperation. Moreover, managing the mechanism involved selecting, developing and teaching workers, processes that psychologists were to claim expertise on. Although it is clear from the historical record that Taylor's mental revolution was to a large extent window dressing, mere talk rather than actual practice, the same is true of other aspects of scientific management. It doesn't change the fact that at least in its theory, its principles, scientific management did formulate a mental side of shop floor management and industrial relations. On the other hand, the human factor that industrial psychologists articulated, beginning with Münsterberg, was as mechanical as the core of scientific management, albeit psychologically more complex, and managing the human factor psychologically was proudly presented as a form of engineering.

The decisive change in management away from the mechanicism at the heart of scientific management was not the rise of psychologists as experts of the human factor but the articulation of freedom and responsibility as the essence of the human factor. Workers were not machines but human beings, not because they had feelings, intelligence or a soul (although they do) but because they would resist being treated as machines, manipulated as cogs in a mechanism. They were free agents, demanding democratic rights and responsibilities.

Governing such self-governing subjects required a new form of management, one with a tactful character. For Taylor, tact had been essential in converting a shop to scientific management: gradually, cautiously, managers and workers had to be persuaded of the advantages of a scientific approach to work. It was a rhetorical process, often involving one-on-one conversations, that aimed to make the introduction of scientific management a collective decision. Once the system was in place, an atmosphere of camaraderie should envelop it, with managers and workers in shoulder-to-shoulder contact on the shop floor. Within the system, however, everyone, including the managers, had to obey its dictates. Tact preceded and surrounded the system but wasn't part of it. In Follett's vision, on the other hand, the manager is the facilitator of a collective process of decision taking. Rather than engineering human beings or giving orders, she has to bring the group to self-control. It is a subtle form of management.

4 'Social Technology'

Introduction

There is a hint of exasperation in the stern admonishment with which Albion W. Small (1854–1926) began his seminars on the foundations of sociology: 'Radical error and persistent confusion would be forestalled, if students could be familiar from the start with the fact that sociology is not, first and foremost, a set of schemes to reform the world.' Sociology is to be 'an accredited section of philosophy' that endeavours to collect all the knowledge that has a bearing on 'the relations of men to each other'.[1] To improve society, one must first know society, its structure and forces. Small therefore first laid out the basic methodological processes – the various forms of analysis, synthesis and abstraction – and only then turned to the question of what purpose sociology serves. 'Why do we want to analyse the world of people?'[2] His answer was threefold: we want to know 'the facts of human experience' and the laws that they are subject to, we want to know what sort of social goals those facts and laws indicate as being reasonable to strive for, and we want to know how to attain those goals. This last question, Small added, 'calls for social technology'.[3]

It was one of the first times the term 'social technology' was used. It appears to stand for a straightforward and plausible idea: the knowledge that a scientific discipline gathers about an area of reality can be applied to intervene in it and further our goals. When we apply a (natural) science in this way, we call this 'technology'; therefore, when we apply a social science we employ a 'social technology'. Yet whereas 'technology' has become one of the keywords of our era, 'social technology' has had a mixed fate.[4] On the one hand, the term 'social technology' has been employed by some of the greatest social scientists of the twentieth century: Albion Small but also Karl Popper[5] (a social thinker if not a social scientist) and Robert Merton,[6] as well as by lesser but still considerable figures such as Dorwin Cartwright[7] and Olaf Helmer.[8] Moreover, authors would use it as if it were a common term for a concept that required no further explanation. So unremarkable did they consider the term that they seldom discuss or refer to its use by others. Thus, Popper writes about 'social technology' without acknowledging the work of Small, Cartwright uses 'social technology' without mention of either Small or Popper, and Helmer in turn ignores all three when he urges the social sciences to start developing 'social technologies'. On the other hand it seems fair to say that 'social technology' has not caught on in the way that for example 'management' or indeed 'technology' have. It never found a place in the vocabulary of social theory. It is, for example, rarely a topic in introductory textbooks in sociology.

The term 'social engineering', in contrast, has had a very different fate, becoming a widely used term in both social science and wider culture. As a rough indication of the popularity of 'social technology' in comparison with 'social engineering', Google's ngram viewer (which searches Google's database of books) shows that apart from a brief period in the first decade of the century, 'social engineering' has always been the more frequently used term.[9] The difference steeply increases in the early 1920s, and while the popularity of both terms varied, 'social engineering' remained the more common term by far. A search of JSTOR's database of journal articles brings out a much finer-grained picture, with an interesting pattern.[10] Between 1895 (when Small founded the *American Journal of Sociology*) and 1920, the term 'social technology' occurs in forty-four articles, of which thirty-five were in the *AJS*. 'Social engineering' occurs in thirty-four articles until 1920 (with nine in the *AJS*), but then its use explodes: 340 articles between 1921 and 1940 and 710 between 1941 and 1960. 'Social technology' meanwhile is found in ninety-one articles between 1921 and 1940 and 103 between 1941 and 1960.

Most remarkable is the fact that 'social engineering' increases tenfold between 1921 and 1940 in all journals but fourfold in sociology journals. 'Social technology' only doubles its numbers in the same period in sociology and other journals. In other words: 'social engineering' shows a remarkable rise in popularity between 1921 and 1940, particularly outside sociology, whereas 'social technology' hangs on in sociology but falls behind elsewhere. In the period between 1941 and 1960, 'social engineering' is by far the more popular term.

Why did 'social technology', such a promising term on the face of it, not catch on in the same way that 'social engineering' did? Why was 'social technology' not adopted to the same degree? 'Social engineering' was apparently the right term at the right moment, but why? And what made 'social technology' unable to fulfil the same role in the vocabulary? What made it awkward or unsuited? To answer this question, I first describe its use in American sociology in the first two decades of the twentieth century, in particular at the University of Chicago sociology department under Small. Next, I chart some of its spread beyond this origin and describe its shifts in meaning along the way. To understand these shifts, I argue, it is necessary first of all to put them in the context of the history of the term 'technology', which I go into in the second half of this chapter.

Focusing on the difference between the terms 'social technology' and 'social engineering' may seem to miss the point. Usually, they are both taken to refer to the rational control and reform of social processes, a scientific alternative to politics. Does it really matter whether this is called social technology or social engineering? Indeed, they were sometimes used interchangeably by the social scientists themselves. Surely, then, what is important is this ideal itself and what came of it in practice. But historical studies have also shown there has been great variety in the way 'rational control of society' was envisioned and realised, and I will argue that some of this variety was captured in the different terms 'social technology' and 'social engineering'. Labels matter: they indicate how the world is classified and how the classifications relate to each other, the connections that people make. Shifts in the vocabulary often indicate cultural transformations. As historian Leo Marx put it, 'keywords often serve as markers, or chronological sign-posts, of subtle, virtually unremarked, yet ultimately far-reaching changes in culture and society'.[11] That 'social technology' never really made it as a keyword and 'social engineering' did is significant.

Small's Sociology

Albion Small was appointed professor of social science in 1892 by William Harper, the first president of the University of Chicago.[12] He had been trained as a Baptist minister but had pursued his interest in social science in Berlin and Leipzig after his graduation, one of many American social scientists to have studied in Germany. On return he took up a teaching post at Colby College in Maine, eventually becoming its head in 1889. At Colby he began to realise his ambitions for a scientific social science, replacing a course on moral philosophy with one on sociology, for which he privately printed a textbook titled *Introduction to a Science of Society*.[13] His appointment in Chicago offered him the opportunity to develop his ideas further, which he did with great success. Although his theoretical contributions were soon forgotten after his retirement, he had a lasting influence with his boundary work, demarcating sociology as a scientific discipline and creating professional institutions for it such as the *American Journal of Sociology*.[14]

As the opening quote suggests, a key objective of his boundary work was to distinguish sociology from efforts at social reform, in particular those of the people united under the flag of the American Social Science Association. Their social science consisted of social surveys, exhaustive descriptive studies of social problems conducted with an eye at intervention. Science and reform were inextricably linked. Against their empiricism and reform aspirations, Small emphasised theoretical knowledge of the laws governing society as a whole. Only on that synthesising basis could intervention in particular areas of society and their problems have a chance of success. Thus, Small asserted the authority of 'theoretical' (or 'general') sociology over 'practical sociology', of professional academics like him over reformers with their 'benevolent amateurishness'.[15]

To his students, Small preached caution regarding the application of sociology to the solution of social problems. In the first handbook of sociology, which he wrote together with George Vincent, they insisted that a science must be wary about practice, particularly in its formative years. Sociology is not 'a collective name for the various schemes by which unscientific optimists expect to organise imperfect men into perfect society'.[16] The book does contain a chapter on social reform, but that ends with a paean to the sociologist as a discoverer of social facts, laws and principles. Men who are capable of such work seldom have the time, or the talent, to do practical work, but nonetheless their importance for the solution of social problems may be immense. Galileo and Newton never bothered with practical work either, but they laid the foundations that made the work of engineers and inventors possible.

In no way did Small reject reform and intervention, however. For Small, sociology was in the end all about social betterment – an 'ethical view of social science' that he had acquired in Germany.[17] He had been particularly impressed by the work of the Verein für Socialpolitik and its founder, the economist Gustav von Schmoller. In his address at the founding meeting of the Verein, in October 1872, Schmoller had made the 'social question' its central concern: its aim should be to solve the problem of the deep division in society, the battle between factory owners and workers, between rich and poor, and avert the looming danger of social revolution.[18] Schmoller reminded his audience that such divisions had brought down the Greek and Roman civilisations of antiquity. The required social reform could not be trusted to emerge spontaneously from the operation of the free market, as the Manchester school of laissez-faire liberals seemed to think, but neither was the social levelling that the socialists advocated the right answer. 'The healthiest, most normal society is one which represents a ladder of different ways of being, but with easy steps from one rung to the next.'[19] The problem is that the middle rungs of the social ladder are missing, while the ladder is stretching and the space between the lower and upper rungs is getting ever greater. Before this widening gap leads to a bitter class struggle that will destroy all free political institutions and bring back absolutism, the state must intervene. The state, however, should be above party politics; it should be the 'greatest moral institute for the education of the human race'.[20] It was to be the role of the economists united in the Verein für Socialpolitik to guide the state in this moral task by drafting legislation.

Like Schmoller, Small believed passionately in the need for social reform guided by social science. In his view, reform efforts should be based on a synthetic theory of the integrated whole of society. This theory, the outline of which he first drew in the 1898 seminar in which he also introduced the term 'social technology', identifies the six human interests that are the root cause of everything people do: health, wealth, sociability, knowledge, beauty and righteousness. In order to satisfy these desires, individuals come together in associations, and the permutations and cross-checking of interests produce ever more complex associations, in what Small called the social process. Because the sociologist reveals the interests that form the foundation of the moral order, he could and should be both an objective scientist and a reformer.[21] Sociology can identify the goals that society needs to strive for, and social technology then gives us the instruments to bring about these goals.

Thus, Small defined sociology as strictly distinct from but at the same time essential for social reform, with the two linked by objectively determined goals and a technology grounded in science. This safeguarded the scientific status of sociology and at the same time offered an alternative to the potentially controversial notion of reform. There was widespread support for reform initiatives, but among the conservative elite that the University of Chicago depended on, reform was another word for radicalism and socialism. Small knew very well that too close an association between sociology and reform could be very damaging to his department and to sociology as a whole.[22] As Small and Vincent stressed in the *Introduction to the Study of Society*, socialism is 'a challenge that society cannot ignore', but the answer is sociology, not revolution.[23] Rather than call for radical reform, it is time to scientifically study the alleged evils of society and develop social programs to alleviate them. Like the Verein für Socialpolitik, Small's sociology promised a way towards the betterment of society that was not political but scientific.

Small once called the Verein 'perhaps the most influential voluntary organization in the world for promoting social technology', but Schmoller himself never used the term 'Sozialtechnologie' or 'Sozialtechnik'.[24] 'Technologie' as such was an unimportant concept in Schmoller's work, in contrast to 'Technik', which in his work 'may be described as a relation between man and nature'.[25] It raises the question of where Small got the term 'social technology'. The term 'Sozialtechnik' did occur in the work of other German scholars. In 1881, a Dr Julius Post published a short tract defending the right to work under the title *Arbeit statt Almosen, Beitrag zur Social-Technik*.[26] More influential was the work of the philosopher Paul Natorp, who, inspired by Plato and Kant, envisioned a social pedagogy that created strong ties between a community and its individual members. Central to this social pedagogy would be a 'soziale Technik', which, like all forms of 'Technik', would allow mastery over its object by exploiting the knowledge of the laws that govern it.[27] However, human beings are determined not only by the laws of nature but also by the goals they set themselves, and 'soziale Technik' must take into account human self-determination and attempt to reach its own goals via those of the people it works on.

A few years before Small, the legal scholar Julius Ofner distinguished the science of law from its application, jurisprudence, which he called a 'soziale Technik'.[28] Ofner espoused a mechanical view of society and the role of jurisprudence in it. Jurisprudence is the art of applying the law, but this art has to take into account not only the logic of the law but also how it works out in practice. Jurisprudence must be sensitive to the unjust consequences that the blind application of the law may have; it must be prudent rather than merely logical. Property, for instance, is the engine of enterprise and a cornerstone of morality, but it is also dangerous where it clashes with the common good and makes slaves out of those without property. It is up to jurisprudence to weigh the advantages and disadvantages of ownership and to navigate between unquestioning devotion to this fundamental social principle and utopian 'social poetry' on the other hand.[29] Jurisprudence must be based on sober, scientific study of the consequences of social arrangements and interventions. It must draw on the study of the laws that govern of human action, which is as constant and predictable 'as the growth of a tree'.[30] Legal institutions like wedlock, property, obligation and inheritance are social machines, and legal reform is analogous to the improvements of physical machines. Thus, jurisprudence has one eye on the law and the other on social justice and brings the two together through interventions based on knowledge of the mechanics of human action. It is a 'wissenschaftliche soziale Technik'.[31]

Ofner's essay is from 1894, and Natorp first used the term 'soziale Technik' in 1895.[32] It is not unlikely that Small had read them both, given his familiarity with the social sciences in the German-speaking world. He refers to neither, however. He would have been uncomfortable with Ofner's mechanicism but would have approved of the rejection of social poetry and applauded the call for a sober, scientific approach to reform. Whether 'Technik' could be translated as 'technology' is a question that I will return to later, but in any case Small seems to have been right at the forefront of the latest developments in social science with his term 'social technology'.

Henderson

After demarcating it as a scientific alternative to radicalism, Small delegated social technology to his colleague Charles Richmond Henderson (1848–1915), who had joined Small at the Department of Sociology in 1892.[33] Small later described their partnership, 'the nucleus' of the department, as a happy marriage of different but complementary characters, with 'mutual reinforcement between men who were primarily interested in the theoretical phases on the one hand, and the applied phases on the other, of sociological knowledge'.[34] Henderson was a Baptist minister with a long career in social reform when he came to the University of Chicago, where he became the university's chaplain as well as a professor of sociology. In contrast to Small, it was more than an idea to Henderson that sociology was a discipline with a role in the improvement of society. He was and remained a visible, widely known figure of the reform movement and held many leading positions in reform organisations, for example, as the president of the National Conference of Social Work in 1899.[35] The sociologist Howard Odum wrote of him: 'He literally bridged the distance between sociology and social work.'[36] At the department, Henderson was put in charge of the courses in applied sociology that attracted the social workers who formed a large part of the students. Henderson published an extensive treatment of the subject in 1901 titled 'The Scope of Social Technology'.[37] If the term 'social technology' was indeed Small's idea, it was Henderson who first defined it and spread it through his own work and that of his students.[38]

Henderson proceeds along the lines set out by Small and first divides the whole of sociology into theoretical and practical social science. Theoretically, sociology is a science of associations that tries to find the laws underlying these phenomena and thereby explain them. Practical social science deals with exactly the same phenomena but with a different purpose, namely 'to discover and present in systematic form principles which regulate social conduct in conformity with ends'.[39] Practical social science encompasses a study of values, the application of these values in the critical evaluation of existing situations and finally 'the problem of the means', or social technology: what should be done and how.[40] 'As theoretical social science culminates in explanation of what is, so practical social science culminates in the best methods discoverable, at a given stage of knowledge, for bringing the actual into approximate conformity with what is required by that situation, required by the inherent facts of the situation.'[41] Like theoretical sociology, social technology is synthetic: the problems it is concerned with involve 'all the elements of welfare' and 'all the members of the community' and require drawing together data from various special sciences. If the problem can be dealt with by 'an expert in any particular field of science or experience, the problem may be left there'.[42] However, the main point is that social technology operates on society as such, not on any particular problem within it.

Developing social technology may be a daunting task, but there is no reason for cynical despair. If we assume that a theoretical social science is possible, then applying the laws it discovers in social technology must also be possible. Of course, specialisation will be necessary, just as it has been in the natural sciences, and Henderson proposes a tentative division of labour based on groups and classes in society, many of which already have a 'body of technical experts' associated with them.[43] Such branches of social technology may include 'domestic science' that will bundle the now hopelessly isolated efforts of school boards, city councils and others who deal with families.[44] And what about a social technology of the rural community? Why do we have so much knowledge about breeding cattle and hardly any about 'forming citizens and organizing culture?'[45] In the distant future we may have social technology of an even larger scope, but 'the social technology of mankind must wait until we test our methodical tools on more limited subjects'.[46]

In its operation, social technology is modelled on science: after formulating the desired end, one carefully lays out all the causal relations in the problem at hand and takes stock of the actions already being taken and their results; next, one constructs a plan (this corresponds to drawing up a hypothesis in theoretical science, Henderson explains) and proceeds to carefully implement it (corresponding to the experiment). Depending on the results, the plan may be adjusted and a new trial made. Henderson stresses that the planning and the trial should be performed by administrators, voters and reformers, or at least in close consultation with people with practical experience, 'the men of technique'.[47] Social technology offers 'general working directions', not 'rules of art made by local administrators for particular direction of specific processes'.[48]

A later exposition of social technology (now a synonym of 'applied sociology') by Henderson underscores its simultaneously moral and scientific character. Referring to Small's classification of fundamental human desires or interests, Henderson explains that the ends to which society is to be modified are objective facts, 'like a star or a crystal'.[49] After analysing the desirable ends, social technology aims to understand under which conditions they may best be realised. For this it may draw on the knowledge and expertise of various sciences, but their insights must be synthesised by the sociologist because '[a]ll forms of science culminate in applied sociology'.[50] It is, after all, society that is to be ameliorated. Ultimately, the goal is to improve the general welfare, and again Henderson makes it clear that this offers the applied sociologist a criterion that is both objective and right: 'so long as any group of humanity is ignored, so long our judgement of a tradition of social conduct must fall short of being scientific as well as ethical'.[51]

From Chicago, the concept of social technology spread to other American sociology departments. In a 1902 survey,[52] Henderson's seminar on 'Methods of Social Technology' was the only sociology course in America using the term, but seven years later a similar survey showed that social technology was a subject widely recognised in American sociology programs.[53] The 1909 survey had been organised by Albion Small and conducted by his student Luther Lee Bernard (1881–1951), and it appeared in the *American Journal of Sociology*, of which Small was editor. No wonder then that social technology was one of the subjects listed in the questionnaire, which was 'prepared under Professor Small's direction and criticism and under the criticism of a number of the leading professors of sociology'.[54] Question 10 asked respondents to mark with a cross the divisions of sociology 'which your plan of instruction recognizes'.[55] Social technology was one of ten subjects listed, together with for example methodology, general sociology and rural sociology. Although Bernard acknowledges encountering some disagreement about the classifications,[56] he reports that social technology is a division of sociology recognised in more than a third of the institutions, more than for example methodology.

The spread of the term 'social technology' over American sociology departments in the 1900s is a sign of the enormous influence of Small's department. The might of Chicago was rivalled only by the sociology department at Columbia, led by Franklin Giddings. Small and Giddings conducted a bitter feud, but they agreed in demarcating sociology from the superficial efforts of the Social Science Association. Sociology was a science because it saw further than the surface of society. Whereas Small had stressed the synthetic, coordinating character of sociology, Giddings claimed that it uncovered the foundations of society, but the upshot was the same: reform must be based on the facts and theories discovered by sociologists.[57] Speaking before the Social Science Association in 1894, Giddings stressed that sociology 'is not philanthropy: it is the scientific groundwork on which a true philanthropy must be built'.[58] In the rest of his address, Giddings explains what makes sociology fundamental, but he says nothing more about its relation to philanthropy. Whereas Small combined the emphasis on science with a firm belief in the ethical role of sociology and gave it separate attention, Giddings didn't think the solution of social problems required more than the scientific method. In 1891, while still a lecturer at Bryn Mawr, he had written that dividing the subject of sociology into theoretical and applied, or theoretical and practical sociology, or science and art, was 'the easy device of incomplete or baffled thinking'.[59]

He repeated the assessment and his arguments (extensively copying verbatim from his article) in *The Principles of Sociology*, the textbook that he wrote after he was hired as professor of sociology at Columbia. Some of the facts that a science deals with are simply more practical than others because they affect our daily lives more, but 'as knowable facts they admit of an explanation'.[60] Thus, a 'scientific arrangement' of social problems will give us a more adequate view of their nature and 'create a scientific order in the maze of facts of practical sociology'.[61] Although Giddings considered sociology and reform to be quite compatible and had been an active supporter of workers' cooperatives,[62] he did not believe in the need for a social technology separate from sociology proper.

Science, Technology and Techniques

In Henderson's description, social technology has a position midway between theoretical sociology and social practice: it is a part of science devoted to assisting efforts at social amelioration, but it is not itself such an effort. Henderson demarcated social technology in the same way that Small demarcated sociology: by stressing its synthetic character. Speaking of domestic science, for example, he notes that 'there is not a science or art which has not made a contribution', but what remains to be done is to draw it all together.[63] 'The materials for a division of social technology lie scattered about, and society sorely needs a systematic coordination and construction of them in a coherent body of regulative principles.'[64]

The same emphasis on general, regulative principles is found in the work of Henderson's student Jesse Steiner, professor of social technology at the University of North Carolina from 1921 to 1927.[65] Social technology is in between sociology and social work, and its role is to bring together 'the principles and data of the social sciences that throw light on social problems and aid in building up social work practice'.[66] 'Social technology becomes a field of its own with its demand for specialists who have a working knowledge of both social science and social work.'[67]

Steiner had moved away from Henderson's grand conception of social technology: he did not envisage a social technology of mankind or even the rational reorganisation of rural communities; rather social technology is simply the application of social science to social work. However, the mediating role of social technology has remained the same: it draws on sociological theory to supply practical work with regulative principles.

Other sociologists, however, gave social technology a much more practical meaning, referring to concrete instruments such as the social survey, 'a unique bit of technology which is capable of estimating and reporting social facts quantitatively without sacrificing the soul of these facts altogether' according to Carl Taylor.[68] Sometimes it was synonymous with applied sociology[69] or social work.[70] Small himself vacillated on the issue. At one point he simply glossed 'social technology' as 'the whole body of approved devices for promoting social progress in every department of life', dispensing with the generalising aspirations of 'principles' in favour of the concrete, almost material 'devices'.[71] In *General Sociology*, the definitive statement of his view of the discipline, he offers 'principles of adapting means to ends in practical improvement of society' as a definition[72] but further on in the same book again drops the principles and settles for 'the ways and means of bringing the social purposes to pass', illustrated by various tactics of political manoeuvring.[73]

These ways and means for improving society were also referred to as 'techniques'. Henderson in 1904 distinguished technique from technology. The latter tries to answer the question what our moral duty towards the poor is, for instance, by calculating an acceptable standard of living. What the best way of fulfilling our duty is, how the welfare of the poor can best be raised to that acceptable standard, is a question of technique. Small made a similar distinction when he saw a task for sociology in '[d]eveloping technologies as distinguished from techniques of social improvement; for instance, restorative v. retributive theories in penology; or vocational v. cultural conceptions in education'.[74] Such technologies are value-laden conceptions of practices: for example, a conception of education as a matter of training for particular vocations, or rather as a process of personal growth and cultivation. Techniques are then the practical ways and means of realising such forms of education.

'Social technology' continued to make regular appearance in the texts of American sociologists, its meaning moving about in the semantic space between science and practice, social work in particular.[75] Mostly it is a term without much weight, used once or twice in a text, often in passing or in the context of some other subject. However, it is also a term that needs little or no explanation, its meaning apparently assumed to be known. Typical for its status is the fact that the term does not occur in

Jesse Steiner's book *Education for Social Work*, although the byline on the title page presents the author as a professor of social technology.[76] After Henderson's 'The Scope of Social Technology', there is little discussion about it, indicating its modest but also secure place in the vocabulary. The only explicit criticism of the term is in a brief discussion by Charles Ellwood. In his own demarcation of sociology, Ellwood, who studied with Small, saw no room for applied sociology. A general science like sociology forms the basis not for one but for many applied sciences, just like biology is the foundation of the various medical sciences and of agriculture and horticulture. We do not speak of 'applied biology', so why would we need a term like 'applied sociology'? One might, he concedes, use applied sociology or social technology as a term for the 'organization of all our knowledge of practical means and methods of improving social conditions' as Henderson had proposed, but it would be better to drop the term altogether.[77] Apart from Ellwood's remark, however, 'social technology' remains an accepted term, used by some, ignored by others.

Technique and Engineering

After 1920, the terms 'technique' and 'technical' become increasingly prominent in the discourse about the practical application of sociology. Characteristic for this development is the *Introduction to the Science of Sociology*, coauthored by Ernest W. Burgess and Robert E. Park, the dominant Chicago sociologists in the 1920s. In their textbook (known as 'the green bible') Burgess and Park do not use the term 'social technology' but 'technique' and 'technical' all the more, for instance, when they discuss 'the technical (applied) social sciences'.[78] Park was largely responsible for creating the new approach to sociology that became known as the Chicago School, characterised by empirical studies of often marginal, inner-city groups such as gang members using participant observation. 'Social technology' was one of the victims of the regime change. When Park taught his first course at Chicago's Department of Social Science, in the academic year 1912–13, 'nearly half of the courses in [the] department were in the subdivision of "Social Technology"'.[79] In 1924, a year before Small left, 'social technology' was dropped from the course listings.[80] One reason the department reverted to the original term 'social institutions' must have been Park's emphasis on science rather than practice.

Park's demarcation of sociology followed the same logic as Small's before him: sociology, in order to be a science, must distance itself from practice. Park's exasperation about those who confused sociology and social reform – he referred to them as 'damned do-gooders' – equalled

Small's own.[81] In contrast to Small, however, Park rejected moralism altogether, preferring 'a scheme of life that should be devoted to merely seeing and knowing the world without any practical aims whatever'.[82] It was a common attitude among the second generation of American sociologists.

Historians have described the early history of American sociology in terms of a struggle between historicist and positivist approaches, between an emphasis on meaning and purpose in describing social processes and one on mechanism and objectivity.[83] Although the discipline was not divided neatly into two camps, and historicism and positivism would sometimes vie for dominance in a single person or be joined in concepts such as Small's objective social goals, these two perspectives on society and on sociology's role in it are broadly recognisable, and the positivist perspective gained the upper hand in the 1920s. Inherent in this perspective was the conviction that values must not interfere in the discovery of facts, that science is amoral and – as Park put it – 'without any practical aims whatever'. With his upfront reform ideals based on a strong Christian faith, Henderson in particular, even more than Small, was increasingly out of place in the new climate. When he died in 1915, Henderson was a symbol of ideas and ideals quickly going out of fashion.[84]

At an informal conference at the eighth annual meeting of the American Sociological Society in 1913, the differences between the generations became clear. The topic of the discussion, suggested by Small, was the question of whether it was 'possible for American sociologists to agree upon a constructive program'.[85] In his opening words, Small repeated his view that it is the task of 'intelligent sociology' to describe life as 'a perpetual problem of the accommodation of human interests', urging his colleagues to '[a]lways do your part toward compelling everyone to treat human experience ... as a connected system of problems in morals'.[86] Moral values are involved in every social problem, and it makes no sense to reject the ethical perspective with the words 'you can't change human nature'. Changing human nature is 'the most prolific industry in the world'.[87] In Germany this is widely understood, witness Professor Schmoller's standing and social influence.

The other speakers at the meeting showed little enthusiasm for Small's views or even for the need to come to agreement about a 'constructive program'. Only Henderson's eminently practical motion to set up a committee to coordinate the field studies of sociology students of different universities was approved by all. In his contribution, Park, who had just arrived at the Chicago department, stressed that '[i]n the last analysis, science ... *is* technique', and this goes for sociology too: it looks for relations of cause and effect in human affairs and lays down rules that allow prediction from one situation to another. On this basis 'all formal methods or organizations for the control or improvement of social life must rest'.[88] Of course, he adds, in practical work it is important to understand individual people, their thoughts and feelings. Some claim this requires a different, historical kind of science, but the same kind of insight is provided by the intensive sociological study of 'typical and individual cases', people who are representative of 'classes, races, nationalities, or localities': the kind of studies the Chicago School would excel in.[89]

Thus, although Park's sociology was incompatible with moralism, it did not preclude practical application but it was now considered in terms of 'technique' rather than 'reform'. In their textbook, Burgess and Park frequently employ mechanistic metaphors for society and its institutions. 'Administrative problems are mainly practical and technical. Most problems of government, of business and social welfare, are technical.'[90] Government, administration and education are machines and devices, and sociology offers the knowledge and technical means to improve their efficiency. 'Science, natural science, is a research for causes, that is to say, for mechanisms, which in turn find application in technical devices, organization, and machinery, in which mankind asserts its control over physical nature and eventually over man himself. Education, in its technical aspects at least, is a device of social control, just as the printing press is an instrument that may be used for the same purpose.'[91]

Such a mechanistic view of man and society had wide currency, and many of Park's contemporaries were much more enthusiastic than he was about the promises of social control that they entailed. The first decades of the twentieth century in the United States, the 1920s and 1930s in particular, have been dubbed the Machine Age, characterised not only by the fascination with machines and devices but also by the tendency to apply the machine as a metaphor for social processes. It was a metaphor that promised the possibility of the rational control and planning of society. As John Jordan has described it in his history of Machine Age ideology, engineering became the alternative for politics, replacing 'social contention with definitive answers'[92] and promising rational and efficient improvement of society. The epitome of the Machine Age was the technocratic movement, but outside this cult-like group the basic idea of society as a machine that could be designed, tended and repaired in the way engineers did with physical machines was widely shared among politicians, journalists and social scientists.[93] Sociologists like Steiner, Odum and William Ogburn (the latter two students of Giddings) were involved with initiatives for rational reform such as Herbert Hoover's

Research Committee on Social Trends.[94] The commonly used term for rational reform, however, was 'social engineering', not 'social technology'. Ogburn, for example, in his presidential address before the American Sociological Association, discussed the application of sociology in terms of 'social engineering' rather than 'social technology'.[95] Although 'social technology' did not disappear completely, it was certainly overshadowed by 'social engineering'.

At this point, we seem to be faced with a paradox. For Small, and for Henderson even more, sociology had a moral task, apparently at odds with a term like 'technology'. Why did Henderson, the Baptist minister with a prominent position in social reform and charity, adopt the term 'technology', with its engineering connotations? Later, with the rise of scientific management and its extension to public administration, 'social technology' should have come into its own, but it didn't. 'Technique' and 'technical' became commonly used terms in connection with social work in particular, but not 'technology'. Applying science to improve society became known primarily as 'social engineering'. The paradox, in other words, is that 'social technology' appeared when the circumstances seemed unfavourable and started to whither when the tide was turning in its favour. To make sense of this it is necessary to broaden the perspective to include the history of the concept of 'technology'.

Useful Arts, Technique, Technology and Technocracy

When Henderson and Small started to use the word 'technology', it was much less common than it is now. In the same year that Small first used it (1898), Henderson published his first book on sociology, *Social Elements*.[96]. In it he doesn't use the term 'technology' but speaks of 'the useful arts' being 'the means and methods by which society secures its satisfactions'.[97] Among the useful arts are those practised in industry, but they also include parenting, preaching and teaching, poetry and agriculture.[98] As historian of technology Ruth Oldenziel has shown, at the time Henderson wrote *Social Elements*, 'the useful arts' was a term on the way out, gradually being exchanged for 'technology'.[99] In the process, the inclusiveness of 'useful arts', which allowed both grocery shopping (one of Henderson's examples) and forging steel in the same category of practical and artful activities, fell victim to something much more restricted. Technology eventually came to mean devices and artefacts designed and tended by (male and middle-class) engineers; in other words, technology became a matter of men and machines. Perhaps agriculture could still count as a technology, but not parenting, and the link between building bridges and writing poetry was broken.

The word 'technology' had already been in use in English, albeit infrequently, since the seventeenth century.[100] Until the middle of the nineteenth century 'technology' referred to the academic *study of* the practical arts. As 'technology' began to replace 'useful arts', however, it came to refer to those arts themselves, 'the artefacts and processes of industrial civilization'.[101] Eric Schatzberg has traced this semantic shift to the influence of the German word 'Technik'.[102] Like its English cognate 'technique' and the ancient Greek *technè* from which they both derived, 'Technik' originally referred to the rules, methods and skills for achieving a certain end (the technique of a pianist, for example). By the mid-nineteenth century 'Technik' was also increasingly used in connection with the industrial arts to denote the material means of production. Thus, 'Technik' had two related meanings: 'a narrower one referring to the material aspects of industry, and a broader one encompassing the rules, procedures, and skills for achieving a specific goal'.[103]

'Technik' in the industrial sense was a popular concept among German social scientists, including Gustav von Schmoller.[104] When Thorstein Veblen, the maverick American social scientist, wanted to discuss the crucial importance of industry in transforming society, he sought a way to translate 'Technik' in this sense. Because the English 'technique' did not refer to material artefacts and had artistic connotations, he chose the term 'technology' instead. Veblen first used 'technology' in this sense in 1901, in a review of Schmoller's magnum opus, *Grundriß der allgemeinen Volkswirtschaftslehre*.[105] In this way, according to Schatzberg, Veblen was instrumental in shifting the meaning of 'technology' from the study of the industrial arts to those arts and artefacts themselves.[106]

Veblen's work was a key influence on the technocracy movement and on Machine Age thought in general, although he always remained something of a fringe figure.[107] In Veblen's work from 1901 onwards, technology represented the productive, beneficial forces in society as opposed to absentee ownership, pecuniary institutions and other parasitic elements. In his late work *The Engineers and the Price System* Veblen decried the 'sabotage' of the production processes by both workers and owners – the wilful obstruction by the workers and the curtailing of production by the owners in order to keep prices high and required labour at a minimum.[108] It was time, Veblen argued, for a 'soviet of technicians' to take control of the industrial system and manage production in a rational fashion to the benefit of all. He had his hopes pinned on the young generation among the engineers, those who hadn't sold out yet to the 'captains of finance' and other 'Vested Interests'.[109]

Veblen's vision of engineers as revolutionary agents met with a largely hostile reception, not least from American engineers themselves, who preferred to present themselves as inventors, constructors and repairmen working in a basically sound social system. Maier has shown that in the controversy over Veblen and the technocracy movement, 'technology' was used rhetorically to counter the claim that technicians should have the power to re-engineer society.[110] Veblen's conservative critics used 'technology' to stand for the consumer goods that industry produced, such as radios and washing machines. 'Technology' in this sense stood for new comforts and welfare rather than for the unemployment that the industrial system had produced according to the technocrats. Thus, whereas in Veblen's use of the term, 'technology' referred primarily to what he also called 'the machine process', his critics emphasised the objects produced by this process. All in all, by the late 1930s the English word 'technology' had largely, but not entirely, shed its meaning of 'the study of the useful arts', instead referring to the machines used in industry and the machines produced by industry. When William Ogburn discussed the relation between sociology and technology in 1938, it was not to talk about social technology but to point out that technological innovations – steam power and contraceptives, for example – lead to changes in social institutions.[111]

This history of the term 'technology' in the first half of the twentieth century may help to resolve the paradox of 'social technology'. To start with, its adoption by Small and Henderson loses its mystery. When Small introduced 'social technology', 'technology' was not strongly associated with material, particularly industrial processes, but rather referred to the study of the principles involved in practical arts, including but not restricted to those in industry. 'Social technology' allowed sociology to claim relevance for, even jurisdiction over, social reform and social work without thereby damaging its claim to the status of science, which was Small's primary objective.[112] As Henderson explained, the actual work was to be done by men with practical experience, the 'men of technique'. The distinction between 'technique' and 'technology' was still well in place when Henderson drew the outlines of social technology: 'technique' is what artists and artisans use; 'technology' is the study of their art. The social technologist is not, in Henderson's scheme, a 'social engineer' getting his hands dirty. 'Technology' implied a comfortable distance from practice. At the same time, it avoided the potentially controversial connotations of moralism, reform and politics, instead suggesting the objectivity of science. 'Technology' lent social work and other forms of intervention the authority of science without risking contamination of science by the unsavoury elements of social reform.

Next, several developments conspired to make this meaning of 'social technology' as a bridge between sociology and social practice obsolete.

As we have seen, the meaning of 'technology' shifted away from the study of, and the principles behind, the application of scientific knowledge to those applications themselves.[113] At the same time, its meaning narrowed to material processes and devices, thus excluding social practices. Developments in sociology and social work further undermined 'social technology'. In Chicago, the theoretical sociology of Small, of which social technology was originally conceived to be the practical application, gave way to the empirical studies of Park, Burgess and their students. Application, let alone 'social technology', was not a focus of the hugely influential Chicago School.

In addition, after World War I, sociology lost ground in the scientific foundation of social work.[114] As social work professionalised, technique was increasingly emphasised as the mark of the professional that distinguished her from the well-meaning volunteer. Technique, however, was learnt in practice, doing fieldwork, not from books and sociology courses. Most professional schools of social work offered 'a concrete, practical curriculum which subordinated theory and research to technique'.[115] The social workers assembled by Thomas Eliot in a virtual 'symposium'[116] on the relation between sociology and social work often used the term 'technical' in connection to their work and training, but only one spoke of 'social technology'.[117] Moreover, social workers in the 1920s increasingly shifted their focus from the social environment of the case to his or her personality and family history. For their scientific foundation, social workers looked to psychology and psychoanalysis rather than sociology. Environmental manipulation gave way to analysis and treatment of the personality that was exposed to the environment.[118]

This leaves the question of why 'social technology' did not become the label for the project of rational social reform in the Machine Age of the 1920s and 1930s when sociology had joined in the mechanistic thinking that was sweeping society. If society was a machine, then 'social technology' seems a good word for its rational reform. An important factor must have been the very popularity of the term 'engineering' and its extension 'social engineering'.[119] The central figure of Machine Age thinking was not the sociologist but the engineer, 'the shining hero of the moment',[120] and his[121] work was primarily called engineering rather than technology. As we have seen, at around the same time 'technology', through the work of Veblen, increasingly came to refer to material, industrial processes and products, whereas 'social technology' was associated with Henderson's explicitly moralistic sociological project.

Oldenziel notes that in the 1920s engineers themselves still preferred 'applied science' over 'technology'.[122] As Ronald Kline has shown, 'applied science' served to underline the scientific basis of engineering and thus allowed scientists to claim cognitive authority over the work of engineers, and engineers to raise themselves professionally above mere artisans.[123] The rise of 'technology' was due to the work of Veblen and the popularity of the technocracy movement. 'About 1930, "technology" became a buzzword incorporating anthropological notions of civilization, engineering professionalism, and machine metaphors.'[124] This could have been the moment for 'social technology', were it not for the controversy surrounding both Veblen and the technocracy movement. Despite a brief period of national popularity, technocracy was rejected by the 'Vested Interests', and it gradually acquired the bad name that, to most people, it still has. Worries over Nazism, Fascism and Communism also took the shine off the ideal of rational reform and wholesale social engineering. Thus, when the term 'technology' really broke through in the 1930s, social engineering was becoming increasingly suspect, and extending the term 'technology' to the goods that industry produced was more attractive than including society in its ambit.

What's in a Name

Although 'social technology' was overshadowed by 'social engineering' after World War I, particularly outside sociology, it did not disappear completely. One example is its use by Beardsley Ruml, who after doing graduate work at the University of Chicago in 1922 became the influential director of the Laura Spelman Rockefeller Memorial Fund. Ruml adopted the term 'social technology' for projects of 'applied research', but most of the ample funds of the memorial were channelled to the basic social scientific research that Ruml thought should underlie social technology.[125] Under Ruml, the Chicago School received more funds than any other institution.[126]

Luther Lee Bernard, mentioned above, also continued to use the term 'social technology'. Although his relationship with Small was deeply ambivalent and he 'positively disliked the ministerial Charles Henderson', Bernard did adopt the concept of objective social goals.[127] For example, in a 1928 essay on standards of living (the same subject that Henderson had written about in 1904), he stresses that it is the job of the sociologist to determine objectively the standard of living, i.e. the level of expenditure required to 'promote the efficiency of the individual or group (family) in the type of socially sanctioned production or living which the individual or group is carrying on or seeks to accomplish'.[128] Society sets the goals; the sociologist discovers them and then calculates – using data from sciences such as physiology – the level of consumption that individuals and groups require to achieve the targets. Next, this standard of living becomes an instrument 'in the hands of the social worker and technologist', who uses it to improve the efficiency of the 'workingman' and his family.[129]

Bernard's sketch of the 'pragmatic and functional world of the social technologist',[130] as well as the occasional use of the term 'social technology' by Ruml and other sociologists in the 1920s and 1930s, raises the question of whether the relative decline of 'social technology' and the rise of 'social engineering' are all that significant. After all, one might argue, what remained constant was the ideal of rational, scientifically based reform and control of society. Whether such attempts to 'produce the social' were called 'social technology' or 'social engineering' might appear of little consequence.[131] It didn't seem significant to Howard Odum, who used 'social engineering', 'human engineering' and 'social technology' interchangeably.[132] All these terms seem to express the idea that human behaviour and/or society must be ordered according to a rational model. It is also clear, however, that beyond this core, the diversity of actors, ideas and practices is too great to bring everything together under one term.

First of all, in the period that I have described above, 'social technology' and 'social engineering' were connected, even if in a relatively loose fashion, with different ideas of science and its application to the betterment of society. Originally, social technology was part of a grand, even Utopian idea according to which sociology had the task of determining the goals that society was striving for, collecting and synthesising the scientific knowledge that was relevant to those goals and translating it into plans for social interventions. This was the religiously inspired, deeply moralistic project of Small and, in particular, Henderson. In Bernard's social technology, the element of objective social goals survived into the 1930s.

In practice, 'social technology' came to be associated in particular with social work, but social workers themselves increasingly preferred techniques, acquired on the job, over social technology handed down from a university. What was later called 'social engineering' shared the same grandeur of the original conception but lacked its moralism and religious basis. Its keywords were successively 'efficiency', 'social control' and 'planning' rather than uplift and reform.[133] Society was a machine rather than a complex of associations between people with certain basic interests. Social engineers were available as 'service intellectuals', offering their expertise to further the normative concerns of others,[134] but for a social engineer the rationality of the means really was an end in itself.[135]

Secondly, although it is difficult to say anything definite about the meaning of 'technology', since it so variable, it does appear to have had a semantic potential that 'engineering' lacks. When the distinction between technique and technology was still firmly in place, technology implied a certain distance from practice: it was the *logos* of *technè*, not the practical deployment of those techniques. In Henderson's original description, social technology is academic work; its application is to be left to the 'men of technique'. Henderson's social technologist devises plans, policies and principles. Later, this distinction is eroded, as is clear from Bernard's term of 'technologist' for social worker, but 'engineering' never implied the same distance from the field. The engineer, although not a workingman, often moved among them and tended to draw pride from that.[136] For Small, deeply committed both to sociology as a science and as an objective guide to social reform, the concept of technology offered a way to connect these two missions while keeping sociology purely scientific. I have noted that Small at times glossed 'social technology' in a way that made it synonymous with techniques, but the fact remains that 'technology' at first carried some of its original, more academic meaning.

'Social technology' was a move in a play of recurrent boundary work between sociology and social practice. Small's foil, social science, was itself an attempt to be scientific in contrast to mere philanthropy, and the same is true of the charity organisation movement of the 1880s.[137] In turn, Small and Henderson's moralistic sociology was rejected in the bid of sociologists like Park and Ogburn to create a *truly* scientific sociology. Thus, successive demarcations of sociology (or social science) each depicted their predecessor as tainted by a sentimental, unscientific attachment to philanthropy, social reform or social work. Ogburn's demarcation of sociology, at least in his presidential address, is finally so stridently scientistic that sociology has been cleansed of any aspiration to be useful for practice. 'Sociology as a science is not interested in making the world a better place in which to live, in encouraging beliefs, in spreading information, in dispensing news, in setting forth impressions of life, in leading the multitudes, or in guiding the ship of state. Science is interested directly in one thing only, to wit, discovering new knowledge.'[138] Sociological knowledge may in fact prove very useful for the social engineer or social worker (the two are implicitly synonymous), and it may even happen that sociologist and social engineer are united in one person, but they are strictly separate functions. Ogburn, in other words, had no use for a term like 'social technology' to label the link between sociology and social practice without collapsing sociology into reform. There is simply no link, no grey area to cover. Social engineering belongs on the other side of the fence.

In sum, 'social technology' and 'social engineering' belong to a family of concepts that express the idea of rational, scientific management of society, but the diversity in this family is significant. Historian Thomas Etzemüller has made a valiant effort to bring everything together in one concept, 'social engineering', but ultimately has to describe it as a 'multilayered formation', the layers of which are variously connected by different actors, resulting in quite diverse ideas and practices.[139] Indiscriminately referring to these different kinds of scientific improvement of society as 'social engineering' makes such nuances disappear from view. 'Social technology' should not be considered as a mere synonym of 'social engineering'; it has, to some extent, its own history.

That history did not end in the late 1930s. As I describe in Chapters 6 and 7, after World War II 'social technology' was resurrected in a culture that was preoccupied with the struggle between democracy and tyranny. In the process, 'social technology' would acquire new meanings: in the work of Karl Popper it became virtually synonymous with social science, to be applied in 'social engineering', and in the work of Kurt Lewin and his students it lost its associations with sociology, becoming an applied branch of social psychology instead. Finally, in recent years the two terms have changed meaning again, becoming associated with new forms of sociality: 'social technology' is now often used to refer to online networking services like Facebook, while 'social engineering' is gaining currency as a term for manipulating people into giving confidential information or access to their computer.

5 Dale Carnegie and the Fine Art of Dealing with People

> The unvarnished truth is that almost every man you meet feels himself superior to you in some way; and a sure way to his heart is to let him realize in some subtle way that you recognize his importance in his little world, and recognize it sincerely.
>
> —Dale Carnegie[1]

It All Seems Like Magic

The way he described them, Dale Carnegie's rules and principles for dealing with people bring the dream of effective and efficient manipulation of human behaviour within reach of everybody. To us, the title may seem a joke, but Carnegie's multi-million-seller *How to Win Friends and Influence People* describes the techniques to do just that. Carnegie assures the readers that the effect of their application is astounding: businesses are saved, lives are turned around. 'It all seems like magic.'[2] Carnegie's principles, it appears, afford a leverage on a par with material technology. Just like the automobile can transform a gallon of petrol into 30 miles of travel, following six rules is enough to make people like you, twelve rules to persuade, and seven to secure a happy marriage. What's more, all these rules and principles derive from three 'fundamental techniques in handling people'.[3] Master these three techniques, and you can conquer life's greatest challenge – how to get along with other people. If you do not want to take Dale Carnegie's word for it, you can turn to the many testimonials from satisfied readers of his books and from alumni of the Carnegie Course,[4] the core of his enterprise. Billionaire Warren Buffett still has his course diploma on his office wall and credits the course for changing his life. Lee Iacocca (former president and CEO of Chrysler), Lyndon B. Johnson (the thirty-sixth president of the United States), and Jerry Rubin (scourge of big business and Washington politicians alike) all did the Carnegie Course. Even Charles Manson's career of evil is testimony to the power of Carnegie's techniques. Manson did the Dale Carnegie course in prison in the 1950s, and although he quit before finishing it, his infamous powers of manipulation were in no small part due to what he learned there, according to his biographer.[5]

Carnegie's courses, his books, and the company he created have been phenomenally successful. The sales figures of *How to Win* ... are a case in point. Virtually from the day it was published, Carnegie's 'practical, working handbook on human relations' was a runaway success.[6] Published in November 1936, it was confidently offered by Simon & Schuster with a money-back guarantee.[7] In January the publisher reported that it was the fastest-selling book it had ever published.[8] By early February it had sold 100 000 copies and was a bestseller all over the USA.[9] At the end of the month, sales were double that amount.[10] When it finally lost the top spot in the *New York Times* bestseller list in April 1938, the publisher reported more than 800 000 copies sold.[11] Currently the number stands at more than 30 million, including translations.

When explaining his success, authors often portray Carnegie as a typical American. His work is put in the context of American cultural currents, and the attraction of his books and courses is thought to lie in the fact that they responded to anxieties and desires that were characteristic of Americans in his time.[12] Chief among those typically American factors behind Carnegie's successful career is New Thought, or positive thinking as it was called by Carnegie's friend, the self-help author Norman Vincent Peale. New Thought emerged in the second half of the nineteenth century as a radically upbeat reaction to the Calvinist gloom of puritanism. Thinking positively and banning all negative thoughts was promised to promote both mental health and material well-being. In the hands of Carnegie's contemporary Napoleon Hill (and recently of Rhonda Byrne, author of *The Secret*) positive thinking acquired a paranormal mind-over-matter character, with the assurance that with sufficient 'visualisation' and similar mental techniques anything that one desires can be made to 'materialise'. To his credit, Carnegie was averse to such fanciful ideas, his frequent reference to the effect of his techniques as 'magic' notwithstanding. He prided himself on his common sense and his Missourian 'show me' attitude. But Carnegie agreed fully with New Thought as a philosophy of mental health: positive thoughts make for a healthy mind, and a positive attitude is effective in social life. One cannot affect the material world directly with the mind, but indirectly confidence, kindness and happiness do bring material rewards.

Historian Warren Susman has called *How to Win* ... 'a key work in any attempt to understand the culture of the period', meaning American culture of the 1930s.[13] The book is typical of the conformism of those years, the emphasis on adjusting the individual to the existing social structures. The instructions for how to succeed by adjusting oneself to the social order were provided by the many 'how-to' books of the era.[14] Two key terms in this conservative trend were 'personality' and 'human relations'. Whereas character, central to the nineteenth-century view of the self, refers to the self in relation to moral demands, personality is about the self as it appears to others. How-to books advised their readers on the ways to develop a sparkling, interesting, pleasant personality and emphasised how important that is in getting ahead in life. Society was a field of human relations, interacting individual personalities, and the ones best adjusted to it succeeded in it. Susman briefly notes that Elton Mayo's human relations approach in industrial psychology provided, like Carnegie, instructions for adjustment, but 'from the perspective not of the worker, aiming to "get ahead", but from that of the manager, anxious to provide an effective and happy work force'.[15] What made personality, human relations and adjustment such focal ideas was the rapid urbanisation and industrialisation of America and the growth of consumer culture. Living in large cities and working in large bureaucratic corporations, Americans increasingly had to deal with strangers, and their success in this environment came to depend on how well they got along with others. In addition, the growing surplus of products – people could now choose between four different detergents – and the increase in disposable income gave rise to a consumer culture in which sales and marketing became companies' central concerns. All in all, the 'fine art of dealing with people', as Carnegie called it, had become the key to success in American society.[16]

As significant as the local circumstances of its origins are in explaining why Carnegie wrote *How to Win ...* and why the book met with such remarkable demand, they cannot explain why Carnegie's success went beyond his immediate context: extending from the America of the 1930s to the America of today and beyond the USA to much of the rest of the world. *How to Win ...* and other Carnegie books are still in print and are regularly updated under the auspices of Carnegie & Associates. They remain bestsellers, typical airport books – I bought my copy of the latest edition of *How to Win ...* at Pierre Elliott Trudeau International. Moreover, within months after publication, *How to Win ...* was translated into other languages. Reactions to *How to Win ...* were much the same elsewhere in the world: glowing testimonials from people who had followed his advice or done the course, enthusiastic book reviews in the newspapers mixed with sarcastic and biting criticism, particularly from intellectuals. It is true that certain aspects of Carnegie and his course, such as the focus on success, were often seen as typically American. Nevertheless, Carnegie's books and his public speaking and leadership courses have found global demand – Dale Carnegie & Associates now offers its various courses in more than 80 countries.

The first, obvious explanation for this lasting, worldwide appeal lies in the fact that key aspects of the context in which Carnegie developed his course and wrote his books have spread over the rest of the world as part of the global cultural and economic dominance of the United States. Large, bureaucratic corporations and a consumer culture are found all over the world. As a result, personality and a positive, can-do attitude are valued in present-day China as much as they were in 1930s America. In this chapter, however, I will pursue another explanation for Carnegie's globalisation: it was enabled not only by the expansion of critical elements of the social context but also by the way Carnegie made himself and his methods mobile. He did this, firstly, by standardising and packaging his course, gradually creating what amounted to a standardised product and a brand, 'Dale Carnegie', but also, secondly, by constructing the problem that he addressed in terms that transcended local circumstances. This aspect of his mobilisation was provided by psychology: by constructing social situations in terms of human relations between psychologically defined individuals and disregarding moral, economical, political or other non-psychological issues, these situations acquired the universality of psychological mechanisms. Carnegie appeals in present-day Indonesia as much as he did in 1936 America because he speaks to a situation that is constructed as universal and of all time. As a solution for this universal problematic Carnegie then offers his art of dealing with people, consisting of 'techniques' to influence the psychological mechanisms underlying human behaviour and relations.

Finally, I will point to a paradoxical element in the way Carnegie made his art a technique that transcended local circumstance: his techniques, he emphasised, must never be used as 'tricks'; they only work if one is sincere.[17] For the art of dealing with people to be effective, the 'artist' must be an integral part of it. His techniques of influence were inspired by psychologists like William James and Walter Dill Scott, but what sets Carnegie apart is his realisation that in human relations, technique must always be made personal to be effective. The paradox is that, at least in Carnegie's approach, the personal is made highly uniform. What appears to be local, particular, individual about Carnegie's techniques – personality – is at the same time universal.

Public Speaking

The advice in *How to Win Friends and Influence People* covers a wide range of situations, from contacting a prospective customer to having breakfast with one's husband, but it all grew out of Carnegie's work training people in public speaking. Dale Breckenridge Carnagey, as his name was from his birth in 1888 until he changed it in the 1920s, took his first steps in acquiring rhetorical mastery in church, where his devout mother often let him recite psalms and religious poetry before the congregation.[18] Later, as an insecure 16-year-old, embarrassed by his family's poverty and his ill-fitting clothes, he set his sights on his school's debating championship as the route to popularity, respect and girls. Though failing dismally at first, through determination and many hours of practice he succeeded: he won the championship, other hopefuls sought his advice, his self-confidence soared, and girls began to speak to him. Public speaking, in other words, was a very practical and personal matter for Dale Carnagey.

The rhetorical training he received at college was equally practical and personal. Rhetoric at the State Teacher's College was an example of the shift from oratory to public speaking that took place in American culture around the turn of the century.[19] The old formalism, with its affected style of declamation, its set pieces, its rules and theory, was giving way to a naturalistic approach that emphasised personal engagement and connecting with the audience. Sincerity, speaking from the heart and not by the book, was the prime directive. To some extent, this contrast between the old oratorical tradition and public speaking is the result of the boundary work of the public speaking advocates. Upon closer inspection, it becomes clear that the new approach borrowed a lot from the old. As Michael Sproule has shown, 'the new-style public speaking book emerged not as an innovation in whole cloth, but rather as the outcome of a sifting through of preexisting material'.[20] Nonetheless, there was a clear shift in emphasis from expression to communication and, correspondingly, from lofty ideals to more earthy goals. Nineteenth-century rhetoric was above all a form of art, which as such might contribute to the betterment of society. Public speaking on the other hand served practical purposes such as success in business and advancing one's career. It was about getting things done, and the primary goal of a public speech was to move the audience to act. Public speaking was about influence.

After a successful stint as a travelling salesman in the Midwest followed by a disappointing semester as a student at New York's American Academy of Dramatic Arts and a gradual slide back into poverty, Carnagey in 1912 eventually found a source of income as an instructor in public speaking at a YMCA. His course was a great success: by the end of 1914 he was teaching the 'Dale Carnagey Course in Public Speaking' at YMCA's in New York, Philadelphia, Baltimore, Newark and Wilmington, and renting office space in Carnegie Hall. His first book followed in 1915, coauthored by Joseph Berg Esenwein, a writer, editor and former educational director of a YMCA in New York. Their book, *The Art of Public Speaking*, was a long and very thorough textbook for a correspondence course in public speaking, somewhat at odds with the practical, result-oriented training that Carnagey was giving at the YMCA. Each chapter in *The Art of Public Speaking* is followed by a number of questions that test theoretical knowledge: 'Describe one method of destroying monotony of thought-presentation.'[21] 'Define tempo.'[22] To further deepen the student's understanding and develop his practical skills, there are exercises, often on the basis of classic speeches. Deliver Lincoln's Gettysburg address ('Fourscore and seven years ago . . .') in a fast tempo, and note the ridiculous effect[23] or read it 'over and over . . . until it burns in your soul', and then deliver it with emotional expression.[24]

This focus on theory and attention to the details of delivery and the use of a corpus of classic speeches as didactic material is in line with Sproule's point that public speaking borrowed heavily from the older oratorical tradition. However, *The Art of Public Speaking* also contains phrases, examples, themes and advice that would become characteristic of Carnegie's later repertoire. The many homely metaphors and similes ('Roast beef is an excellent dish, but it would be terrible as an exclusive diet') and the numerous references to the sayings and deeds of famous people (Napoleon, Emerson, Jesus) are certainly reminiscent of Carnegie's style in *How to Win* . . . and other books.[25] Moreover, the main themes of Carnegie's later work are already present. Firstly, proficiency at public speaking is a matter of confidence, and confidence can only be acquired by doing. 'Practise, practise, PRACTISE in speaking before an audience will tend to remove all fear of audiences, just as practise in swimming will lead to confidence and facility in the water.'[26] Secondly, in public speaking, as in any other venture, it is essential to be positive. Only with a positive attitude can the speaker conquer the hearts and minds of his audience and control his own. Thirdly, despite all the techniques and the exercises, persuasiveness cannot be 'slipped on like a smoking jacket'.[27] To connect with your audience, to make them listen and consider your words, you must really believe in the cause you advocate. Sincerity and enthusiasm are key. And finally, suggestion is a more powerful method of persuasion than argument because people are emotional rather than rational creatures. It's here that Carnagey most showed his debt to psychology.

Influence and Suggestion

As we have seen in Chapter 3, when psychologists first entered the terrain of business, they advertised their expertise by arguing that applied psychology stands to experimental psychology as engineering stands to physics. Just like engineers had applied knowledge of the natural world to create the many marvellous inventions that had transformed the Western world in the nineteenth century, so psychologists were ready to apply their knowledge of the mind to engineer behaviour and the social world. But while industry had enthusiastically adopted the products of the natural sciences, it remained sadly ignorant of the riches of psychology. Among those riches, psychologists particularly liked to point to a law that served well to illustrate the engineering potential of their discipline: the law of ideomotor action.[28] Willliam James had formulated it in this way: 'We may then lay it down for certain that every representation of a movement awakens in some degree the actual movement which is its object; and awakens it in a maximum degree whenever it is not kept from so doing by an antagonistic representation present simultaneously to the mind.'[29] James went on to explain that ideomotor action is the default mechanism behind our behaviour. Thoughts lead to action all by themselves – they do not require an added effort of will. What may prevent us from blindly following our impulses is the fact that they are held in check by antagonistic impulses. The resulting tug of war is what we call deliberation. When we hypnotise someone, on the other hand, we prevent this process of deliberation because hypnosis narrows the attention so that antagonistic ideas do not arise to consciousness, leaving whatever idea we implant in the mind free to come to action. Following the ideas of the Nancy school of Hyppolite Bernheim, James and other psychologists considered hypnosis to be a special case of the phenomenon of suggestion, whereby ideas presented to the mind are accepted by the subject unawares, automatically.[30]

To psychologists more interested in the practical application of their discipline than James was, the law of ideomotor action and the technique of suggestion implied a way to engineer human behaviour. In one of his early forays into industrial psychology, Hugo Münsterberg pointed out that psychological knowledge of suggestion, 'from slight therapeutic influence to the deepest hypnotic control', had begun to trickle into the medical world, but it was not yet applied in advertising and sales.[31] 'And yet it is evident that every man behind the counter and every sales-girl who wants to influence the customer works with psychological agencies.'[32] Selling a product to a customer requires, among other things, the 'careful avoidance of opposing suggestions'.[33] In an example that would later also turn up in Carnegie's work: 'Whether shop-girls in a department store are advised to ask after every sale: "Do you want to take it with you?" or instructed to ask first: "Do you want to have it sent to your home?" makes no difference to the feeling of the customers. They are inconscious sufferers from the suggestion, but for the store it may mean a difference of thousands for the delivery service.'[34]

In his book-length exploration of 'the economic psychotechnics of the future', Münsterberg again saw stimulating the will to buy and suppressing opposing tendencies as the essence of salesmanship, but he had to admit he hadn't found 'even a modest beginning' of experimental research on the subject, although there was some work on advertising.[35] As yet, salesmen do their work in ignorance of the psychological laws that operate in it. The world of commerce is still ruled by chance because 'no one has taken the trouble to examine scientifically the psychotechnics of selling and to put it on a firm psychological foundation'.[36] The waste in energy and resources is enormous. And again Münsterberg used the example of the department store, illustrating the savings that could be made by giving commerce a solid psychotechnical foundation. Münsterberg added that a psychotechnics of buying would give consumers some protection against getting trapped by the schemes of the salesmen. 'The salesman must know how to use arguments and suggestions and how to make them effective, but the customer too must know how to see through a misleading argument and how to resist mere suggestion.'[37]

Among the few experimental studies of the effect of advertising that Münsterberg could point to were those of Walter Dill Scott, who had published his first book on the subject in 1903. Eight years later he applied the psychology of suggestion to the more encompassing issue of handling people in general. *Influencing Men in Business; the Psychology of Argument and Suggestion* starts with the customary reproach that the world of business had eagerly gathered the fruits of the physical sciences but had neglected the mental side of production and commerce. Yet, he added, businessmen do know that the ability to handle and influence men is more important than skill in handling things. In an argument that would become a stock part of Carnegie's books and lectures: 'the man who has unusual ability in handling men is sure to attain the position of superintendent or manager; but . . . the man who has great cunning in handling material things is not thereby assured of a position above that of a skilled mechanic'.[38]

Although Scott promises to carefully weigh the merits of the two means of influence, argument and suggestion, it is clear from the outset that he is convinced that suggestion is a far more powerful force than reason. It all comes down to the law of ideomotor action. Common sense tells us that man is a logical, reasoning being, but in truth 'he is a creature who rarely reasons at all . . . our most important acts and our most sacred conceptions are reached by means of the merest suggestion'.[39] In the hands of a skilful businessman, and when applied to people who are susceptible to it, suggestion is remarkably effective. Although there certainly are situations in which argument is called for – when dealing with savvy professional buyers, for example, or when 'selling strictly scientific goods' – in many cases suggestion works better.[40] It is always to be used as a complement to logical reasoning, and it is the preferred method of persuasion when dealing with the general public rather than professional buyers. Rational deliberation, after all, is exhausting, and the 'general purchaser – the housewife – does not ordinarily rise to such an undertaking'.[41]

The influence of Scott's psychology – including the examples and phrases he used to articulate it – is clearly visible in Carnagey's work. His authority is invoked right at the beginning of *The Art of Public Speaking*: 'Prof. Walter Dill Scott says: "Success or failure in business is caused more by mental attitude even than by mental capacity."'[42] The idea that the mind is a mechanism that can be influenced by suggestion lay behind many of the techniques Carnagey advised, and he drew on Scott for it. *The Art of Public Speaking* contains an entire chapter on influencing by suggestion, based on Scott's treatment of the matter.

Remarkably, Carnagey never mentions that Scott had also written about the same subject that he was teaching.[43] In 1906, Scott had published *The Psychology of Public Speaking*, which he adapted for use in schools and colleges twenty years later.[44] Comparing Scott's book with Carnagey's works on the same subject brings out clearly what distinguishes Carnagey from the psychologists that he drew many of his ideas from. To all intents and purposes, *The Psychology of Public Speaking* is a book about psychology, not about public speaking. At the end of each chapter, after an extensive explanation and discussion of psychological theories and experiments, Scott draws a lesson for public speaking, but it seems an afterthought. His heart is not in it, as Dale Carnegie would have said. For example, Scott puts great store in William James's idea that one can create an emotion in oneself by mimicking its expression: 'The voluntary production of the symptoms of an emotion tends to make the emotion a reality.'[45] After a long and detailed discussion of the James-Lange theory, of Cicero and Quintilian on the rhetor's use of emotions and of the testimony of actors on the issue, Scott finally concludes that to move others, one must be moved oneself, and this can be brought about voluntarily. The same principle is central to Carnagey's work – illustrated with the same line from Shakespeare: 'Assume a virtue if you have it not.'[46] Carnagey, however, focuses not on the psychological mechanism but on its use in public speaking, and what is primarily an intellectual puzzle for Scott – do you really feel the emotion that you produce in yourself in this way? – is an aspect of the most important practical issue in Carnagey's advice: the question of sincerity.

What Scott's book lacks most of all is a rhetoric adequate to the task of moving readers to do as the text says. It doesn't employ the pedagogical methods needed to develop men into the kind of personalities required for successful public speaking. *The Psychology of Public Speaking* is a scholarly exposition of the psychology of persuasion, but it does not follow its own principles. It is closely argued rather than suggestive, theoretical rather than practical. Although it was not yet apparent in the book he wrote with Esenwein, Carnagey was already developing the teaching technology to go with the new ideas about public speaking. The core of that technology was Carnagey himself. John Ramage has noted that in later years Carnegie 'maintained an enormous staff of professionals to oversee and run his seminars, produce his course materials, and market his program', effectively constituting something akin to what he calls the 'present day guru distribution system', in which business gurus create stand-ins of themselves in the form of 'an army of consultants, corporate trainers, and human resource specialists who purchase and reproduce the gurus' techniques, seminars, tapes, and books'.[47] The centre of that distribution system, however, and the kernel from which the rest eventually grew, was Carnegie himself. Dale Carnagey employed his whole person, including his life history (properly mythologised), his body, his mind and his words, as a device for producing confident public speakers and, later, generally successful people. This, I think, is the innovation that distinguishes Carnagey from the psychologists from which he drew much of his inspiration. Many of the principles of his art of dealing with people had already been formulated by psychologists such as William James and Walter Dill Scott. What Carnagey added was a system for the effective delivery of these ideas so that they could be put into practice. We can see that system, the trainer Carnagey, develop in the books he wrote after *The Art of Public Speaking*.

Man-Building

In 1920, Dale Carnagey wrote an instructor's manual for his YMCA courses. He was no longer able to meet demand by teaching all the courses himself and had begun to train assistants, often former students of his courses. The manual codified his method and made sure that courses given under his name would meet his standards. The YMCA manual makes clear the at once practical and personal nature of the Carnagey method. It is not about oratory; it isn't even about public speaking as such: what the students want most of all is to have self-confidence, as a survey conducted among students of the course had shown. Students do not want to learn oratorical theory; they want to be able to speak in public without breaking down. The best way to build self-confidence is practice. 'The only way to learn to swim is by jumping in the water.'[48]

In each of the sixteen sessions, students deliver a three-minute talk in front of the class. The instructor gives immediate feedback on every talk, but he gives few lectures and teaches little theory. The students are encouraged to organise their own additional weekly practice classes in order to increase their speaking time. The aim is to get the students to speak in public with 'enlarged naturalness': the spontaneity of ordinary conversation adapted to a larger audience.[49] Instead of oratorical theory and rules of delivery, there are drills and exercises. To practice spontaneity and abandon, the class does the alphabet drill, which consists of reciting the alphabet 'with punch and color', using gestures and facial expressions 'as if the alphabet were a matter of grave concern'.[50] A particularly popular drill is the 'drill on pronunciation', in which the whole class stands and repeats in unison a list of commonly mispronounced words.[51] Halfway through the meeting the class does 'two or three minutes of snappy setting-up exercises' to refresh the students.[52] During the exercises, the windows must absolutely be opened – '[n]o man can generate self-confidence and spirit in a vitiated atmosphere'.[53] Students who are 'below par physically' should be advised to read *How to Live* by Irving Fisher and Eugene Fisk, an all-round practical guide for healthy living.[54] (The rules contained in *How to Live* are surprisingly similar to current consensus, although its advice to sleep outdoors as much as possible now seems quaint.)

In keeping with the practical nature of the course, the manual describes in detail the logistics of teaching, much of it centring on the students' numbers. Every student is assigned a number in order of enrolment. This alone has the effect of persuading more students to enrol because they will want to secure their own number. The instructor enters the number into a 'large, loose-leaf notebook' next to the student's name.[55] On the opposite page, he pastes a letter, written by the student himself, stating his education, profession and reasons for joining the course. During the session, when each student is called forward in numerical order to deliver his three-minute speech, the instructor consults his notebook to see what the aims of this student are and what progress he has made. After each speech, the instructor gives a brief, constructive criticism, and the next number immediately takes his place on the platform. To speed up the proceedings, behind the speaker five chairs are positioned where the next five students sit, waiting for their number to be called.

Secondly, the course is not only practical but also intensely personal. '[P]ersonality is the biggest factor and is the quality which this course aims above all to develop.'[56] For that reason, the second part of each lesson is devoted to what Carnagey called 'man-building'. Personal development happened to be one of the aims of the YMCA as well, and in a passage that must have pleased the association, Carnagey explains that Christian character is an 'invaluable asset' in public speaking, while conversely the training in public speaking develops Christian character.[57] As Carnagey argued, not unreasonably: 'The average man has occasion to speak in public but seldom; but quick, clear, logical thinking, precision in utterance, and self-confidence are assets in daily demand.'[58] Building personality is obviously complicated, but this too has a practical side. A regular reading of the ninety-first psalm ('Whoever dwells in the shelter of the Most High') will do much to inspire confidence, but, after our Lord, the instructor is the most important source of inspiration. His dress must be neat, his appearance successful. He must radiate spirit, enthusiasm and earnestness so that he can communicate these qualities, so essential to public speaking, by 'contagion'.[59] It is, on the other hand, not necessary or even desirable that he is a great orator nor that he has extensive theoretical knowledge. He should be a good public speaker, but all he needs to know is contained in the manual and in the recommended literature.

The students' personalities are put to the test in the tenth, 'heckler's session', in which speakers must try to silence a loud and hostile audience. In the fourteenth session awaits an even more confronting experience, when each student has to stand before the class while the others tell him exactly what his strengths and weaknesses are. All in the interest of increasing the students' ability to influence others and talk convincingly. In sum, the Dale Carnagey Course was a curious mix of the prosaic and the poetic. Its methods and goals were pragmatic, but at the same time they put an almost sentimental stress on the personal and psychological aspects of public speaking. The sixteen sessions, with their roll calls, drills and exercises, employed the familiar technology of discipline, but they were meant to teach the student, ultimately, that to speak well is to be yourself and let go. 'Nothing is more important in effective speaking than genuine sincerity, human sympathy, and love.'[60]

To get this message across, Carnagey instructed his instructors how to teach, indicated which drills and exercises to include and sometimes prescribed with which words to introduce them. In addition, two course books contained carefully selected famous speeches and four special projects (prepare an address for a given situation, conduct a debate on a suitable topic like the League of Nations, etc.). All in all, Carnagey's own voice, if he wasn't himself the instructor, only came through in the descriptions of these projects. This changed when he wrote his own textbook for the course.

A Portable Trainer

That textbook, *Public Speaking; a Practical Course for Business Men*, was the first book he published under the name 'Carnegie'.[61] Changing his name from Carnagey (with the accent on the second syllable) to Carnegie set up a felicitous association with the revered industrialist and philanthropist Andrew Carnegie and at the same time put an end to the persistent confusion over how to pronounce 'Carnagey'. The name 'Dale Carnegie' became a crucial part of the brand. So important was the name to Carnegie that when his niece came to New York to work with him, she too went by the name of Carnegie. On April 13, 1937, the *New York Times* reported that Carnegie had announced the engagement of 'Miss J. D. Carnegie' ('She is Niece of Author') to Gerard B. Nolan. Dale must have insisted that she use his spelling. Ironically, when in the 1930s Carnegie became increasingly famous, other public speaking trainers attempted to capitalise on his fame by also renting offices at Carnegie Hall, just like he had.[62]

With *Public Speaking; a Practical Course for Business Men*, Dale Carnegie instituted himself as the central device of the method. It made his voice an 'immutable mobile',[63] easily transportable to every course session that he couldn't personally attend. In the pages of this book, Carnegie is standing at the shoulder of every student in all of his courses, encouraging him to speak, urging him to stay positive and impressing upon him with his own enthusiasm and fervour that to persuade is to speak sincerely from the heart. Carnegie addresses the reader in a light, conversational tone, with frequent question-and-answer passages. No longer encumbered by Esenwein's writing, he piles on the folksy metaphors and similes, emphasising the practical and common sense nature of the advice. Throughout the book, the tone remains relentlessly upbeat and optimistic, in keeping with the positive thinking credo that optimism gets results. Anecdotes about famous people still litter the book, but they are supplemented with inspirational stories from Carnegie's course and from his own life. For example, the message that you must 'think success' in order to succeed is underscored by a story about Carnegie and a friend scaling a difficult peak in the Alps, armed with little more than self-confidence.[64]

Apart from making the trainer Dale Carnegie portable, *Public Speaking; a Practical Course for Business Men* further codified the Carnegie approach to public speaking into more or less its definitive form. *Public Speaking* contains sixteen chapters in the form of lectures that were to be read in preparation for each session of the course. Confidence is, again, the first subject: students of the course want to gain confidence above all. Carnegie offers a barrage of anecdotes to illustrate the fact that anyone can become a confident speaker but that even the greatest (Churchill, Lincoln, etc.) feel a bit nervous before they get on stage. Finally, Carnegie formulates four principles that summarise his ideas about self-mastery. To become confident, one must start with a strong desire, always be prepared and know what you're talking about, act confident if you aren't yet and, of course, 'Practise! Practise! Practise!'[65]

The first four lectures are devoted to confidence, preparation and the improvement of memory, in other words, to the student and speaker, and only in the fifth does Carnegie turn to public speaking, when he discusses how to keep the audience awake. Even then the speaker remains the primary subject. *Public Speaking* concentrates firmly on building speakers rather than on constructing speeches. Successful speaking depends not on the rules of oratory but on the personality of the speaker. Oratory is dismissed in no uncertain terms: its rules of delivery are 'nonsense and twaddle', a useless waste of time.[66] They are '[a]rtificial, time-killing, mechanical, injurious'.[67] Yet '[t]he average student wants foolproof rules. Something definite. Something he can put his hands on.'[68] And there are such rules, Carnegie admits, but they don't work. What does work is the opposite of rule-governed, mechanical behaviour: naturalness and spontaneity. 'An ounce of spontaneity is worth a ton of rules.'[69] This is the message that Carnegie repeats throughout the course: spontaneity works.

It is, nevertheless, wise to keep certain guidelines in mind when addressing an audience. The traditional rules for gestures that oratory teaches can safely be ignored, but it is all the same better not to make 'jerky movements from the elbow' or repeatedly use the same gesture.[70] Similarly, intuition is the best guide for how to end a speech, but Carnegie does offer some suggestions – a summary of your main points, a biblical quotation or a poem. In general, although we must not mechanically follow rules and become automatons, it does take some effort to achieve naturalness. We must 'practice to be natural before an audience', learn to be ourselves when we speak in public.[71]

Similarly, our personality is our greatest asset in public speaking, and although it cannot be changed much, it is possible to strengthen it or at least get the most out of it by making sure we are dressed for the occasion, for example, and by smiling. Much of the advice on 'platform presence and personality' concerns the physique of the speaker.[72] Make sure you are rested, eat sparingly, and be energetic. Stand straight and don't fidget; and if you *must* sit, do it well: 'A man who knows how to sit feels the chair strike the back of his legs, and, with his body easily erect from head to hips, he *sinks* into it with his body under perfect control.'[73] The most important part of the speaker's body, however, is his voice. Each of the sixteen chapters is accompanied by voice exercises, based on the advice of the great singers and actors of his day, such as Caruso and Madame Melba. Learn to breathe from the diaphragm, like a baby, exercise your lips and tongue, develop resonance, and learn to relax your throat and jaw. The aim is 'enlarged naturalness': we must learn to speak the way we do normally in conversation but in such a volume that the audience can hear us.[74]

The psychological side of public speaking revolves, of course, around influence. As psychology has taught us, it is not reason but suggestion that sways the audience. Scott's law of suggestion – that every idea which enters the mind is held to be true unless it is contradicted by another idea – is presented as 'a psychological discovery of tremendous import'.[75] But Carnegie now also emphasises a crucial precondition of influence, namely tact. Before we can begin the work of convincing people, of winning them over to our way of thinking and getting them to do what we would like them to do, we must first capture our audience. Carnegie illustrates what he means with the example of John D. Rockefeller Jr.'s speech to the striking workers at his Colorado Fuel and Iron Company. Facing a hostile crowd, he had begun by saying that he spoke to them as a friend, having just visited many of their homes and met their representatives, and would now like to discuss their common interests. That, Carnegie enthused, was 'tact – supreme tact'.[76] After this speech, the strikers never mentioned higher wages again.[77] Although Carnegie relies mostly on examples to make clear what tact is – he quotes Mark Antony's funeral oration from Shakespeare's *Julius Caesar* in full – he does offer a few principles.[78] Do not antagonise your audience by starting with your own opinions, but find common ground first. Don't argue, but gently lead your listeners to the desired conclusion. First and foremost: always begin by considering the other's interests and opinions. Here we can follow the simple guideline that people are interested primarily in themselves. '[R]emember that the people you are to talk to spend most of their time when they are not concerned with the problems of business, in thinking about and justifying and glorifying themselves.'[79] By the time he wrote *How to Win . . .*, this simple principle had moved to the core of Carnegie's philosophy, crucial not only in business and public speaking but in all of life.

Influencing Human Behaviour

Carnegie's advice had always had a larger scope than just public speaking. From the start, his course aimed to build better men, not just better speakers: men with the confidence to face an audience, the presence of mind to collect their thoughts and construct an argument on the spot, and the personality to raise interest in what they had to say. Moreover, these were skills that were required not only in business but in other areas of life as well. They were 'assets in daily demand'.[80] Carnagey boasted in his manual that the course attracted business owners and managers, doctors, dentists, lawyers, clerks, engineers, foremen and labourers. Even women followed the course.[81] Everyone aspired to persuasiveness and influence. In the 1920s and 1930s, Carnegie increasingly began to address the goals of his course as universal challenges of modern life. For example, the penultimate chapter of *Public Speaking* opens with a question: if you could have any one of your talents doubled or tripled in power, which one would you choose? The answer is obvious, according to Carnegie: you would choose to have your power to influence others increased. This art, after all, is essential to all of us. Our success in life depends on it. The ability 'to get action' is crucial not only in the world of commerce but in any other area of life as well, whether at home, in the office or in the marketplace.[82]

This widening of his perspective was given a boost when Carnegie read *Influencing Human Behavior* by the philosopher Harry A. Overstreet. So enthusiastic was he about this book that he contacted Overstreet and asked him to lecture on this subject in his courses.[83] At the time, Overstreet (1875–1970) was the head of the Department of Philosophy and Psychology at the City University of New York. He was also, more importantly in this context, a lecturer at the New School of Social Research, which had been founded in 1919 as an institution for adult education with a strongly progressive, free-thinking character. *Influencing Human Behavior* contained the lectures that Overstreet had given in a course at the New School. Overstreet introduced its subject as 'the central concern of our lives ... whether we be teachers, writers, parents, merchants, statesmen, preachers, or any other of the thousand and one types into which civilization has divided us'.[84] That problem is 'to be, in some worthwhile manner, effective within our human environment' – Overstreet lacked Carnegie's gift for memorable phrases.[85] Our entire social life revolves around the process of influencing people, including ourselves. It is 'the supreme art',[86] the 'art of all arts'.[87] Too often the problem of influence is approached purely from a moral angle, but honesty and high ideals are not enough to solve it. If we want to 'become skilled artists in the enterprise of life', we will have to consider influence from a technical standpoint and inquire *how* we can change people's behaviour.[88]

Thus, Overstreet made the move that had remained implicit in Carnegie's work up to that point: to consider all of social life, not just public speaking, as a technical problem in human relations. Its solution lies in a turn to psychology, which in this case is a combination of elements from behaviourism and from psychoanalysis. Behaviourism teaches us about the basic mechanism of behaviour: habit formation through conditioning. Psychoanalysis tells us that this mechanism is motivated by drives. To influence others, we must appeal to these drives: the basic desires of human beings. The effective persuader starts by taking the perspective of the other and considering what *he* wants rather than speaking from his own standpoint and interests. '[F]irst arouse in the other an eager want.'[89] For Overstreet this approach to persuasion was also a matter of respect for individual freedom. For instance, when we train children with what he called 'the older technique of duty',[90] we effectively use compulsion. When, on the other hand, we appeal to their own wants to persuade them to do the right thing, we do not give commands but open up opportunities. Or, in a phrase that sums up what Nikolas Rose has called the 'obligation to be free': we replace the external discipline of duty with 'the subtler, but far more powerful discipline of an inner choice of that which is really worth choosing'.[91]

As the book progresses, the discussion moves from the basic techniques involved in influencing people to a critique of current society and a view of the kind of social life that we could build with these techniques. The mind-numbing discipline and rote learning in schools, the subordinate position of women and the self-interested competitiveness of business all receive trenchant criticism. The 'heaven concept' is described as the epitome of self-deception and a hindrance to social progress.[92] If we continue to diagnose such phenomena and the social problems they cause as religious, moral or political issues, to be solved with sermons, punishment or laws, they will remain insurmountable. Only if we learn to see individuals as habit systems and society as their aggregate will we enable ourselves to treat society's ills with the techniques of influence that psychology offers. Such radical social criticism was not Dale Carnegie's style, and he must have found Overstreet's critique of religion particularly hard to stomach. To the extent that he envisioned 'a better society' it remained low key and secondary to his focus on a better individual. Reading Overstreet, however, certainly deepened his appreciation of the social nature of public speaking, the importance of 'human relations' and the effectiveness of a psychological approach.

Friends and Influence

In his introduction to *How to Win Friends and Influence People*, Dale Carnegie relates how a committee wishing to organise a course on how to get along with people had once asked Harry Overstreet whether he knew of a suitable textbook about this subject. Overstreet had answered that there was no such book, although his own came close. To fill that gap, Carnegie added, he had decided to write his 'handbook on human relations'.[93] Comparing the two books, one can see that the differences between them are indeed slight but significant. To Overstreet's treatment of the subject Carnegie added his inimitable style, a steady focus on personal success rather than a better society, and an emphasis on the importance of friendship, not just influence. In fact, the first thing to note about *How to Win* ... is that the part about winning friends precedes the one about influencing people. Business, and life in general, is not primarily a rational, moral or legal matter, conducted through argumentation, it is a psychological process conducted through human relations. It is therefore necessary to learn how to make friends before one can begin to acquire the skills to influence people. In other words, the 'and' in the title means 'in order to'. The point is made clear in the title of one of the French editions: *Comment se faire des amis pour réussir dans la vie.*[94]

It all starts with the basic principle that one shouldn't judge other people but try to understand them. It is a principle that is at once scientific, moral and practical. To approach people in this way is both 'profitable and intriguing', and it 'breeds sympathy, tolerance, and kindness' to boot.[95] Then follow two closely related fundamental techniques. The first, 'the big secret of dealing with people', is that you can only get people to do something by making them *want* to do it.[96] Barring the use of physical force, the only way ('yes, just one way') to move people is to appeal to their own desires.[97] In Overstreet's hands, this principle became a cornerstone of social reform, but Carnegie turns it into the precept that in dealing with people, we must 'give honest, sincere appreciation'.[98] When we do, we appeal to the most basic desire of all, the desire to feel important. Having thus primed our audience, we can apply the third technique and get people to do what we want by talking about what they want. To underscore the importance of this rule, Carnegie quotes Overstreet: 'He who can do this has the whole world with him. He who cannot walks a lonely way.'[99] Having laid the groundwork with these three basic techniques, Carnegie offers 'six ways to make people like you', which include smiling and remembering people's names and, again, making them feel important and talking in terms of their interests.[100] Only after this thorough introduction to the art of making friends are we ready to consider the 'twelve ways to win people to your way of thinking'[101] and the 'nine ways to change people without giving offense or arousing resentment'.[102] Influence rests on a basis of friendly relations.

Winning people to your way of thinking requires a good deal of self-sacrifice. We will have to renounce our own inclinations in order to win the other over. Although some of the advice in this part of the book is straightforwardly technical – dramatise your ideas, appeal to noble motives – much of it revolves around the self-restraint that influence demands. Never get into an argument, and never tell the other he's wrong. Let him do most of the talking, and make him feel your idea is his. Try to see things from his perspective, and be sympathetic with his ideas. Such techniques are matters of tact. Again and again Carnegie points out the need to operate 'tactfully', 'gently and tactfully' or 'adroitly', and he praises the tact and diplomacy of Benjamin Franklin, Abraham Lincoln and Charles Schwab.[103] For example, one should avoid getting into an argument, but if you really must argue, 'don't let anybody know it. Do it so subtly, so adroitly that no one will feel that you are doing it.'[104] The necessity of tact and self-restraint follows from the basic principle that people are self-centred. '[Y]ou are interested in what you want. You are eternally interested in it. But no one else is. The rest of us are just like you: we are interested in what we want.'[105] Given this somewhat bleak psychological reality, to persuade others we must do the unnatural thing and rise above ourselves. Everyone is only interested in themselves, but (therefore) you must be interested in others. The elitism this implies is sometimes subtle, sometimes evident, as in Carnegie's advice 'to be above fools'[106] or in his quoting of Carlyle's dictum that 'a great man shows his greatness by the way he treats little men'.[107]

Another reason why we must operate with tact is that human beings resist being manipulated. Like Overstreet, Carnegie saw people as self-centred but also freedom-loving individuals. For that reason they resent being told what to do. 'No man likes to feel that he is being told something or told to do a thing.'[108] This is why suggestion is such an important technique in influencing people, why we should 'let the other fellow think the idea is his' and why it is better to ask questions than to give orders.[109] But human recalcitrance also forces us to be sincere in dealing with people. It is, for example, not enough to feign interest in others: we must be unnaturally, but sincerely interested in others. Much of *How to Win ...* revolves around the issue of sincerity. The problem first arises in relation to the second fundamental technique: that of making people feel important by giving them appreciation. Carnegie foresees his readers' response: 'Bear oil! Flattery! I've tried that stuff. It doesn't work – not with intelligent people.'[110] Carnegie agrees: flattery seldom works and usually does more harm than good. Only sincere appreciation will do the trick, and to attain sincerity, we must transform ourselves. 'No! No! No! I am not suggesting flattery! Far from it. I am talking about a new way of life. Let me repeat. *I am talking about a new way of life.*'[111]

In order to be sincere in your appreciation, you must become a person who sincerely appreciates. The recipe is the same in all cases. Time and again Carnegie warns against using his techniques as simple tricks: the smile must come from the heart, the interest must be real. And if you cannot find it in yourself, then it is up to you to change that. Carnegie offers little concrete advice on how to 'perfect yourself'.[112] He recommends reading Henry Link's *The Return to Religion* for advice on how to 'develop a more pleasing personality'.[113] If you still don't feel like smiling, simply force yourself. After all, William James tells us that emotion follows outward expression. But mostly, Carnegie simply sticks to a paradoxical command: 'Become genuinely interested in other people'.[114] 'Make the other person feel important – and do it sincerely.'[115]

What Carnegie does offer is a host of examples of just the kind of personal transformation that is required to influence people. Ramage criticises Carnegie for resorting 'to pure exhortation when it comes to the one thing technique cannot do – render his version of a suitable self sincere'.[116] I do not think this is quite correct: Carnegie does exhort, but he also offers a plethora of examples and himself leads by example. *How to Win* ... is full of little biographies and anecdotes that illustrate the principles of influence and their beneficial, magical effect. Some people, like Charles Schwab and Franklin D. Roosevelt, have a natural talent for dealing with people, but others had to learn their lesson the hard way. Benjamin Franklin, 'tactless in his youth', was warned by a Quaker friend that he would soon have no friends left.[117] He resolved to stop criticising and became so skilled at dealing with people that he was made ambassador to France. Young Abraham Lincoln revelled in ridiculing people and pointing out their mistakes, until one of his victims challenged him for a duel; it was called off at the last minute, and Lincoln never wrote an insulting letter again. Others learnt their lesson in the Carnegie Course. William B. Steinhardt, a New York stockbroker and a terrible grouch, was assigned to give a talk about smiling and decided to try it for a week. It revolutionised his life: he became 'a totally different man, a happier man, a richer man, richer in friendships and happiness'.[118]

Carnegie's own life is a rich source of edifying and inspirational stories. As he confesses in the preface, over the years he had come to realise that he needed training in dealing with people just as much as his own students, and *How to Win* ... contains some of the lessons he learnt: how he had to be told that you shouldn't prove someone wrong and make him lose face just because you know better, how the kids in his neighbourhood only stopped making fires in the park after he stopped telling them to and instead approached them with friendly understanding (he liked a good fire as much as anyone). *How to Win* ... starts with an introduction by Carnegie's friend, the adventurer and media personality Lowell Thomas, in which he describes Dale Carnegie's life as an epic rise from rags to riches.[119] Carnegie's own career is painted in vivid colours as 'a striking example of what a man can accomplish when he is obsessed with an original idea and afire with enthusiasm'.[120] Thus, the possibility of self-transformation is illustrated right from the start with the example of Dale Carnegie himself. Carnegie's career is in a sense the original event that inspires all the life-changing experiences of the students in his course. The road to friendship and influence begins with a transformation of the self, and behind that lies the transformation of Dale Carnegie, his magical rise from abject poverty in Missouri to wealth and global success.

How to Win ... is a technical manual for a successful life. Everything about it, from the title to the layout, speaks of technical efficiency. Its advice comes in the form of numbered lists of techniques, rules and principles, each printed in bold. Their efficacy and power are attested to by the fame, fortune and success of those who practised them. These techniques are not 'mere theories or guesswork' – they are the result of 'fifteen years of experiment and research'.[121] They work. At the same time, on closer inspection most of the techniques that Carnegie offers appear to be composed of self-transformation and tact. They are tactful in that they take the other into consideration – his wants, his viewpoint, his sensitivities – and they require self-transformation so that they may be wielded sincerely, not as mere tricks. The social world is composed of self-interested, freedom-loving, somewhat recalcitrant individuals, but if one operates tactfully yet sincerely in it one can still win friends and influence people without inciting resistance. Carnegie's own life is the best example of that.

With *How to Win* ... Carnegie made a decisive step in the process of increasing the reach of his methods: their applicability was now, in principle at least, universal. Wherever people entered into human relations (thus: everywhere), Carnegie's fine art of dealing with people could be employed to win friends and influence people. The road to success starts with the insight that people everywhere are ruled by the laws of psychology. Understanding people this way, and not judging them by moral or other standards, unlocks the treasures of influence. I have also shown, however, that this movement towards a universal scope was not just theoretical but also involved codifying his training methods and developing a persona as an inspirational trainer. The YMCA manual made the first step in a continuing process of defining what a public speaking course must consist of in order to be called a 'Carnegie Course'. A further move followed with the publication of a textbook for the course, making not just his methods but also his voice mobile.

In the process, Dale Breckenridge Carnagey became 'Dale Carnegie', a brand built around the remarkable person with his mythical history. After Carnegie's death in 1955, his second wife, Dorothy Vanderpool Carnegie, would further solidify and extend this brand. With her keen business sense, something that he had lacked, she built Dale Carnegie & Associates into a global company. Carnegie's approach, in sum, is both universal in its scope and technical in nature as well as highly personal, with Dale Carnegie himself as the central figure. This dual aspect of the Carnegie way is reproduced in the student, who, as Carnegie tirelessly insisted, must be sincere in his use of the techniques of influence. The Carnegie approach demands a self-transformation of his student, requiring each student to become a bit like Dale.

Sincerity

Something of a Carnegie craze swept over America after the publication of *How to Win ...* Carnegie toured the country and spoke before packed auditoria. He became a sought-after speaker at meetings of organisations. Newspapers featured articles about and interviews with the famous author. Charities sought his endorsement, and companies like New York's fashion store Wallach's bought the right to use his name in their advertisements. A 'Dale Carnegie game' appeared in the stores: 'He told you how to win friends – now he shows you how to entertain them!'[122] Carnegie became a household name, synonymous with good relations and making friends. Jokes played on the contrast between Carnegie's friendliness and Hitler's aggression.

But not everybody was impressed. A *New Yorker* reporter went out to observe the 'friends-winning movement' first-hand.[123] He visited one of the Carnegie Course's free introductory meetings and came back with a rather condescending piece about the eager but pathetic human beings dragged onto the stage by the instructor, who effusively praised each attempt at a speech, no matter how poor. Carnegie's earnest enthusiasm was also an easy target for humorists. Irving Dart Tressler took aim at the upbeat message of *How to Win ...* with his parody *How to Lose Friends and Alienate People*. Tressler closely mimicked the style and structure of Carnegie's book – including the anecdotes, the similes, the rules and principles, and a dedication to 'a man who doesn't need to read it' (Hitler, in his case) – to dispense advice to misanthropes. 'Start an argument and make an enemy. Make an enemy and lose a friend. Lose a friend and gain some time to yourself.'[124] *How to Lose . . .* briefly joined *How to Win . . .* on the bestseller lists, and movie production company Metro bought the rights to both books, intending to combine them into a comedy.[125] The plan never came to fruition.

More serious reviewers often criticised Carnegie for valuing appearance over truth and substance. The *New York Times* admitted that 'the principles he sets forth are sound enough'. It is the assumptions behind them that are bad: that one should be enthusiastic no matter what the object of enthusiasm is and that 'the superficial cultivation of personality may take the place of – or even be more important than – a sound foundation of knowledge, intelligence, and ability'.[126] Others were more scathing. James Thurber treated the book with disdain, quoting a few rules and chapter titles ('Never Tell a Man He Is Wrong', 'How to Make People Like You Instantly', etc.) and adding only 'I think that gives you the idea'. He limited his explicit criticism to Carnegie's liberal use of exclamation marks and his claim that one can be sincere and strategic at the same time. 'Unfortunately, the disingenuities in his set of rules and in his case histories stand out like ghosts at a banquet.'[127] The prominent minister John Haynes Holmes accused Carnegie of making a mockery of true friendship. He simply ignored Carnegie's insistence on sincerity. 'The thought that one should flatter, cajole and lie to win a friend is disreputable.' Carnegie's new way of life comes down largely to the pursuit of financial success, to be achieved by liberally spreading praise. It is no match 'for either the Old or New Testament's way of life'.[128]

Novelist and Nobel Laureate Sinclair Lewis delivered the most biting critique. In his weekly column in *Newsweek* he ridiculed the string of titles under Carnegie's name (B.Pd., B.C.S., F.R.G.S., Litt.D.), his adoration of Lincoln and his florid prose ('Great God Almighty!'), but through all the mirth his disgust was evident. Like the bourgeois depicted in Lewis's 1922 novel *Babbitt*, Carnegie advocated nothing but conformism and the pursuit of material gains – his frequent use of 'the magic expression "million dollars"' was a case in point.[129] What Carnegie taught the Babbits of this world was simply 'how to smile and bob and pretend to be interested in people's hobbies precisely so that you may screw things out of them'.[130] Moreover, the methods he proposed were deceptive manipulations, the stock in trade of the con man. A week later, Lewis devoted his column to the man who for him was the antithesis of Carnegie, Henry Thoreau. On the shores of Walden Pond Thoreau had realised, almost a century earlier, a simple yet dignified existence, guided by his own principles rather than by the expectations of society. Thoreau had renounced exactly the kind of 'mean and sneaking' life that was now advocated by the 'Bard of Babbitry', the kind of life that required one to lie and flatter in the hope of getting ahead, a life ruled by avarice rather than by a love of truth and beauty.[131]

The charge that Carnegie's sincerity is false sticks to him to this day. In her study of the self-help genre, Micki McGee calls the Dale Carnegie smile reassuring but disingenuous.[132] Others see psychological or even logical problems in Carnegie's advice. Barbara Ehrenreich remarks how hard it must be to 'fake sincerity' and cites a study by sociologist Arlie Hochschild that showed that flight attendants become depressed because they have to be cheerful all the time, losing touch with their own emotions.[133] Donald Meyer has argued that the injunction to smile sincerely is inevitably self-defeating. Manipulating yourself into being sincere in your appreciation of other people leads to an infinite regress, 'an endless effort to disarm oneself as other'.[134] John Ramage identifies a similar paradox. Carnegie's basic technique comes down to 'endlessly converting self-erasure into power over others in a kind of weird psychic version of cold fusion'.[135] What if everyone were to behave this way? 'In a world full of Carnegian opportunists falling all over each other in an effort to be more interested in the other than the other is in them, no one can be influenced or manipulated by anyone else, no matter how egregiously obsequious.'[136]

I doubt whether such quasi-logical reasoning can decide on the viability of Carnegie's methods, but it is clear that his signature combination of strategy and sincerity makes many people deeply uncomfortable. Perhaps more worrying than the possibility that the smiles are fake is that they might be real: that some people really have sacrificed their former self to become both sincerely appreciative of others and strategically successful, and that for these people there is no contradiction between sincerity and strategy. Maybe the problem is not that this combination is logically impossible or a hazard to mental health but that it is taboo. The Carnegie smile mixes categories that many of us prefer to keep distinct. It is both a technique that is deployed to ascend the social ladder as well as an expression of heartfelt kindness, and it only works as a technique because it comes from the heart. The world of technology seems to have infiltrated matters of the heart and turned friendship into a tool. Personal relations have become prey to instrumental thinking, and at the same time the instruments have been personalised: they are techniques, but it is a soft technology, one that works because it is spontaneous, heartfelt and human. For people like Sinclair Lewis, however, true humanity lies precisely in refusing to play the game of social advancement and rejecting the instrumental thinking that it requires. Being fully human, living life with 'dignity', is incompatible with the quest for success and, most of all, with making your own self an instrument in that quest.[137] Transforming yourself to better play the game is tantamount to losing your individuality.

Carnegie's *How to Win* . . . has been criticised on many points, including its fascination with financial success and adoration of the rich and famous and its complete disregard for social, political and economic circumstances. Ever since its publication, however, Carnegie's treatment of sincerity in *How to Win* . . . has drawn the most ire. Behind the criticism that the Carnegie smile and his other techniques of influence are insincere, I suggest, lies the idea that technology is not human. Lewis's disgust is representative of a revulsion that Carnegie's techniques and principles arouse in many who read his book: a revulsion provoked by the violation of a taboo. The self must not be an instrument because, although people use machines and devices, technology is essentially different from us. Erasing the distinction between self and instrument, sincerity and strategy, is wrong. The claim that Carnegie's sincerity is fake is not so much an empirical observation or a logical deduction, it is a normative judgement.

Epilogue

> The minute he smiled the way that was supposed to be sincere, and, in the voice that was intended to be winning, said, What line of work are you in, Mr. Sloan? I knew he was one who'd read that book and wanted to make friends and influence people, so I thought I'd let him practice his instructions while I amused myself in my own anti-social way.[138]

In William Saroyan's short story 'The Dale Carnegie Friend' the narrator, Mr. Sloan, describes how he plays a prank on a young man who accosts him at a party. Sloan, quietly enjoying his Scotch while ignoring the rest of the party, has his pleasant reverie disturbed by a man of 'about thirty-four, a graduate of Stanford, and a natural-born half-wit with none of the charm of the true, uneducated and unambitious half-wit'.[139] The young man had obviously read Carnegie's book and was keen to put its advice into practice and make a friend. 'You meet an average of three people like him at every party you go to. They used to be alright, but that book is wreaking terrible havoc on them all over the place.'[140] Sloan leads his hapless victim to believe that he is in the gambling business, then that he spent time in San Quentin, next that he killed three people and finally that he has only six months left to live, testing the young man's ability to smile and stay positive to breaking point. Finally the Dale Carnegie friend is saved by the hostess, who whisks him away to meet other people.

The irony is that Sloan combats the young man's attempt at sincerity with deception and that in deceiving him it is actually Sloan who loses. The world that Carnegie has created, Saroyan seems to be saying, is a murky mass of smiles and positivity, where people can't tell the difference between a cock-and-bull story and the truth. 'The more ridiculous the story got the more I tried to let him know I was only having fun, but he didn't get it.'[141] Together with true individuality, the truth as such has disappeared, or has at least become irrelevant. In human affairs, appearance is all that matters. Sloan may have succeeded in fooling the Dale Carnegie friend, but that only proves that Carnegie rules. The ending of the story drives home the point. After sending the young man off, Miss Avery, the hostess, inquires what the two had been talking about:

He's been telling me about recent advances in the control and elevation of mass behaviour, I said.

Oh, you brilliant young men, Miss Avery said. Come along now, you must meet everybody.[142]

6 Karl Popper's Social Technology and the Personal Element

During the 1930s, social engineering began to pickup negative associations.[1] Once the hope of a progressive, rational society, it now began to smack of Soviet-style state planning, totalitarian tyranny and other forms of contempt for individual freedom. Moreover, the technocrats' version of rational reform seemed irrationally radical, and programs of social engineering came with much bureaucracy. By the late 1930s, few believed the promises of social engineering any longer, and many distrusted the very idea. At the same time, new, psychological forms of engineering arose that were employed in the service of profit rather than a better society. Rather than reform society's institutions along rational lines, these techniques worked on the irrational, unconscious parts of the individual psyche to influence and mould preferences, choices and relations. '[A]dvertisers, who applied psychological insights to masses of individual consumers, were soon infinitely more important than the social theorists who attempted to rationalize America. Those who sold things via new mass media successfully exploited human irrationality while social engineers worked futilely to make society logical.'[2]

Luther Lee Bernard, unreserved champion of social engineering in the 1920s (see Chapter 4), sounded the alarm in 1935. The best-developed social technologies, he noted ruefully, are those that are employed for material purposes – banking, advertising, business organisation – and they are commonly used moreover 'for antisocial rather than for proper humanitarian ends'.[3] The application of sociology for base motives such as profit without an eye for the greater good, or even for grand but evil and oppressive schemes, was an affront to the true calling of sociology, but our technologies for crime control, good government, city planning, and 'the abolition of poverty' remain woefully underdeveloped.[4]

Bernard urgently advised a two-pronged response. Firstly, social scientists should get the opportunity, through 'schools or other educational systems', to inform the public about the misuse of social technologies by 'sinister forces' in society.[5] Secondly, the social sciences need to double their efforts to develop social technologies for humanitarian ends, backed up of course by a scientific identification of what those 'wise social ends and objectives' really are – in this respect, Bernard remained faithful to Albion Small's moralistic view of social science.[6] We must not be led by received wisdom in building a better society, 'all ideals, to be of any value in practice, must be built upon the logic of science'.[7] And once those rational ideals have been determined, people can be made to accept them through education, with which he meant behavioural conditioning.

The war that followed confirmed Bernard's worst fears. The advances in psychological science and technology, which could have pointed the way towards a better world, had been put to use in psychological warfare and propaganda as well as for making profit off gullible consumers. The little progress that had been made in the social sciences and their technologies was similarly employed for the wrong ends: for organising 'industrial exploiters' rather than workers and consumers, for example, and 'almost everywhere social technology finds much greater employment in the repression of popular movements toward freedom and human welfare than in their realization'.[8] The war had made painfully clear that Man's efforts to 'conquer his moral world' were a dismal failure compared to his successful conquest of physical nature.[9] The solution remained the same: better education of the public and a form of government that is led by social science rather than by politics. '[A]bove all, we need to develop social technologies looking forward to human welfare which equal those already created for the control of the physical and biological worlds.'[10]

In his reflection on the war, Bernard, then 75 years old, sounded weary and bitter, but other, younger scholars were already working on the question how science could best be employed in the service of democracy. In the United States, Kurt Lewin and his students had developed techniques to foster democratic processes in groups, techniques that would later be labelled a 'social technology'. Their work will be discussed in the next chapter. In New Zealand, the philosopher Karl Popper had worked out a conception of social engineering in which science, technology and democracy were virtually indistinguishable aspects of the same process. For both Popper and Lewin, totalitarianism and the experiences of World War II formed the background of their thinking. Both were looking for ways to bring about a democratic, free and rational society, but they knew very well that social science could be and had been employed for evil and totalitarian ends. But how could society be guided in the direction of democracy in a way that was itself democratic and not coercive? How could people be directed to freedom? For Popper, the answer included a strict prohibition against moulding individuals and social relations. For Lewin, however, that was exactly where the solution lay.

Popper's Piecemeal Social Engineering

As a young Viennese philosopher, Karl Popper had made a name for himself in the international community of philosophers with his critique of the logical positivism of the Wiener Kreis.[11] His *Logik der Forschung*, published in 1934, travelled along the scholarly network of the Wiener Kreis, reaching beyond Vienna and the small community of specialist philosophers. Popper began to correspond with scientists, mathematicians and philosophers in Europe, England and the USA. At the same time, with the rise of Nazism the social and political situation in Vienna was becoming increasingly precarious. As an academic of Jewish descent (although baptised as a Lutheran), Popper was particularly affected by the rampant anti-intellectualism and anti-Semitism in Vienna. He harboured no illusions about the future, having been convinced for years that the social democratic parties were powerless to resist the rise of totalitarianism and that Germany would annex Austria and start a new war. With the situation in Vienna quickly deteriorating in 1935, Popper used his network to plan an escape. Eventually he was appointed at Canterbury University College in Christchurch, New Zealand, where he would spend the war years, far away from Nazi Germany, although uncomfortably close to the encroaching Japanese army.

By this time, Popper's ideas had begun to broaden from a focus on epistemology to include political philosophy. The core of his critique of logical positivism – falsification instead of induction as the key to progress in science and a corresponding emphasis on open, critical dialogue – was also applicable to social progress and the problem of government.[12] In New Zealand, Popper began work on two texts in which he worked out his ideas about political philosophy: the essay 'The Poverty of Historicism' and the longer work *The Open Society and Its Enemies*. I shall focus here on the former because it has more to say about social science.

'The Poverty' is known for being somewhat inconsistent; it was written over a period of eight years, during which Popper's intentions and ideas gradually changed, and before it was published in the journal *Economica*, the editor, Friedrich Hayek, made substantial cuts in the text, which further undermined the clarity.[13] Nevertheless, the ideas about social engineering that it contains are clear enough. Popper develops a democratic conception of social engineering, one in which the rational control of society does not imply a commitment to Marxism or other forms of autocratic rule. Popper acknowledges straight away that the term 'social engineering' may arouse suspicion because of its association with 'collectivist planners, or perhaps even ... the "technocrats"'.[14] The key is to distinguish between 'Utopian' and 'piecemeal' social engineering. The Utopian engineer starts from a vision of a perfect society and devises an all-encompassing plan to create it, often starting by wiping the slate clean. Violence is a common instrument of Utopians, and their belief in their own vision often makes them blind to any failures of the measures they impose. Social disasters like Nazi Germany and Soviet Russia are the result. The solution is not a laissez-faire liberalism but piecemeal social engineering: a cautious 'tinkering' with social institutions in exactly the same way that scientists approach the world, carefully breaking a problem down into its components, setting up experiments and monitoring their outcomes and, most importantly, having an open discussion about every aspect of the process. It is this modest, self-critical attitude that most distinguishes piecemeal from Utopian social engineering.

Popper's main target of criticism in 'The Poverty of Historicism' is a type of social science – historicism – that maintains that the sciences of society are necessarily different from natural science. According to Popper, historicists like Karl Mannheim claim that social processes are essentially historical and that for that reason many of the principles and methods that guide the natural sciences cannot be applied to the study of society: experiment and quantification are impossible, regularities cannot be generalised beyond the particular culture and period in which they are found, and short-term predictions are always inexact. Instead, historicists aim at an intuitive understanding of social events, and rather than attempting to predict their exact course in the short term, they offer long-term, large-scale forecasts of the course of history: historical prophecies, in other words. Such prophecies tell us about a future that we cannot prevent from happening; we can only prepare for it. The predictions of the natural sciences, on the other hand, tell us what to do if we want to bring something about: they are 'technological predictions'.[15]

Popper then proceeds with a critique of the historicists' claims, taking the demands of practice as his touchstone, because '[t]he more fruitful methodological debates are always inspired by certain practical problems which face the research worker; and nearly all methodological debates which are not so inspired are characterised by that atmosphere of futile subtlety which has brought methodology into disrepute with the practical research worker'.[16] The upshot of the discussion is that the historicists are wrong to champion large-scale forecasts (prophecies) as the only possible goal of social sciences like economics and sociology: the social sciences are about technological prediction, just like the natural sciences. In this essay at least, Popper seems to consider social science as essentially a technology: social technology.[17] 'The social sciences have developed very largely through the criticism of proposals for social improvements, or more precisely, by investigations designed to ascertain whether or not a certain political or economic action is likely to produce an expected or desired result. This approach, which might indeed be called the classical, is what I have in mind when speaking of the technological approach to social science, or of "piecemeal social technology".'[18] This doesn't mean, he adds, that theoretical problems are unimportant: the technological approach is so fruitful also because it tends to give rise to theoretical problems. Most importantly, the demands of practice impose discipline and keep us from idle metaphysical speculation. The social sciences should not be looking for their own Newton or Darwin but for a Galileo or Pasteur: someone with a practical, experimental approach rather than a grand, unifying theoretical vision.[19] Popper's vision of social institutions as machines and social science as technology is reminiscent of the ideas of Park and Burgess, discussed in Chapter 4, although they preferred the word 'technique'.

Piecemeal social technology produces 'sociological laws or hypotheses' that are analogous to the laws of the natural sciences.[20] Popper gives a few examples: in a centrally planned society, you cannot have a competitive price system; political reform always leads to repercussions that are undesirable from the perspective of the ends aimed at; if you give people power, they will be tempted to misuse it. To what extent these hypotheses are supported by evidence is another matter, but the point is that piecemeal social technology is no different from natural technology. Next, piecemeal social *engineering* is the application of this technological knowledge – as far as I'm aware, Popper was the first to distinguish social technology and social engineering so clearly and explicitly. Social engineering is exactly analogous to physical engineering; it just builds social mechanisms – institutions – instead of physical machines. 'Just as the main task of the physical engineer is to design machines and to remodel and service them, so that of the piecemeal social engineer is to design social institutions, and to reconstruct and run those already in existence.'[21] Popper uses the term 'social institution' in the widest possible sense, covering both public institutions like law courts and schools, and private bodies such as shops and companies. To the technologist, they are all means to ends, instruments for attaining goals; he looks at them as 'machines rather than as organisms', as the historicist would do.[22] What ends these machines serve is emphatically not the concern of the social technologist and social engineer: 'Piecemeal social engineering resembles physical engineering in regarding such ends as beyond the province of technology. (All that technology may say about ends is, at most, whether or not they are compatible, or realisable.)'[23]

Popper's exclusion of all normative considerations from social science and technology is an important difference with the ideas of Albion Small, Charles Henderson and Luther Lee Bernard. Like Popper, they had defined 'social technology' as academic work, a science of practice, but for them, this technology should serve to realise the goals of society that sociology has objectively determined. To Popper, social science *is* mostly social technology; there is no moral science behind it that gives it direction. The objective determination of society's aims is a historicist fantasy. Science can only say which goals are realisable, not which ones are right.

When he first developed his political philosophy in New Zealand, Popper had only a very small college library at his disposal in addition to the few books he had been able to bring.[24] This may explain the fact that Popper appears to have been completely unaware of the American social technologists and social engineers. Malachi Hacohen has argued that Popper's foray into political philosophy instead bears the traces of Viennese progressivism, an influential political movement in early-twentieth-century Vienna. 'Militantly secular, politically radical, trusting in social reform, popular education, and technological progress, these progressives represented the young Popper's social and intellectual milieu.'[25] Joseph Popper, a distant relative, was the most influential intellectual among Viennese progressives. This Popper, who published under the pen name of Lynkeus, expected the solution of the 'social question' from a scientific management of society: 'He saw no conflict between his deep commitment to individual rights and technological control. Sozialtechnik, or social engineering, became the foundation of fin-de-siecle progressive utopias.'[26] According to Hacohen, Karl Popper spent 'much of his life refashioning progressive philosophy and politics'.[27]

However, Popper gave no indication that he was aware of Lynkeus's use of the term 'Sozialtechnik', or anyone else's, for that matter. When he introduces the term 'social engineering' in *The Open Society*, Popper refers to a book published in 1940, but he adds that he only read it after completing the text for his own book, thus implying that the term was his own independent invention. The term 'social technology' was suggested to him by his closest collaborator in New Zealand, the economist Colin Simkin.[28] In later years, Popper would add more references for social engineering, as he gradually became aware of its history.[29] 'Social technology' remained without a history, however, Simkin remaining the sole reference. Hacohen concludes that Popper had apparently forgotten Joseph Popper and the Viennese progressives, and that he ignored or didn't know about Otto Neurath's early Utopian publications, in which he used the term 'Gesellschaftstechnik'.[30] As Hacohen writes: 'Simkin's "social technology" should have evoked vivid memories. It didn't.'[31]

The Personal Element

Another difference with the ideas of Small, Henderson and Bernard, is that Popper explicitly and emphatically rejects psychological engineering. After setting out his mechanical view of social institutions, Popper adds that he doesn't deny that there are 'fundamental differences between institutions and physical instruments'.[32] On the contrary, it is the task of the social technologist to study these differences and then formulate them as hypotheses with the same technological form as his other hypotheses. The example that Popper then gives concerns the 'personal element': 'You cannot construct foolproof institutions, that is to say, institutions whose functioning does not very largely depend upon the persons involved; at best, they can reduce the uncertainty of the personal element, by assisting those who work for the aims for which the institutions are designed, and on whose personal initiative and knowledge success largely depends.'[33] Because of this personal element, 'the efficiency of institutional "machines" is limited'.[34] Utopian social engineers, in their hubris, do not accept this limitation and attempt to deal with the problem by moulding the people to fit the institutions that they want to build. Rather than creating a society fit for people to live in, they engineer people that fit their idea of a good society.

In the course of the essay, Popper offers three loosely connected arguments against such control of the personal element. The first is that if one intervenes both in the institutions of society and at the same time in the people who live in it, there is no criterion by which to gauge the success of the experiment. We deprive ourselves of 'any possibility of testing the success or failure of the new structure': people who do not like the new institutions just need further moulding to fit in.[35] Thus, Utopian social technology can never be falsified; it is immune to criticism. For social engineering to work as an experiment, it must be measured by a variable that is not itself being manipulated in the experiment, namely the satisfaction of the public. Engineering the personal element makes social experiments unscientific.

A few pages later, Popper returns to the issue, when he claims that Utopians (Karl Mannheim in particular) believe it is possible 'to mould personal intercourse'.[36] Society consists of social relations, and Utopians think it is possible and necessary to control these relations. Now Popper's argument is 'logical': moulding social relations would fail, 'if only because with every new control of social relations we create a host of new social relations to be controlled. That is to say, the impossibility is a logical one.'[37] Popper seems to say that social control leads to an infinite regress because control is itself a social relation (between controller and controlled) and brings other social relations into being, for example, when it becomes institutionalised.

Finally, in the conclusion of his essay Popper returns to the personal element, calling it '*the* ultimately uncertain and irrational element in our theory',[38] and then advances the argument that controlling it will lead to tyranny, subjecting the people 'to the whim of one man, or of a few men'[39] – another paradox, one might say. But suppose we let science – 'the opposite of whim' – control the human factor?[40] Popper agrees that this may become possible: 'In theory biology and psychology probably can solve, or soon will be able to solve, the "problem of transforming man".'[41] But to do so would backfire on science itself because science depends on the 'free competition of thought, i.e. upon freedom'.[42] We must never tamper with the human factor, irrational as it may be, because on it depends our hope for the future. 'If the growth of reason is to continue, and human rationality to survive, then the diversity of individuals and their opinions, aims, and purposes must never be interfered with (except in extreme cases where political freedom is endangered). Even the emotionally satisfying appeal for a common purpose, however excellent, is an appeal for the end of rational thought.'[43] Social technology, in other words, is a technology of social institutions, never of people and their relations. What's more, this technology, and science and democracy as a whole, depends on the rejection of human engineering. Science, technology and democracy require the freedom of the individual, and to engineer the individual is to constrain that freedom.

Leaders

Popper's attitude towards psychological matters can perhaps best be described as detached. He was particularly critical about what he called psychologism, the idea that the laws that govern social processes ultimately derive from human nature and that society is therefore explicable in psychological terms. The only thing that is right about psychologism is its insistence on methodological individualism: society consists of individuals and their actions and interactions – and nothing else. Popper agrees that families, teams, companies, nations or any other kinds of social whole have no reality over and above the people that make them up and how they interact with each other. But that doesn't mean that we can explain social phenomena as the result of purely psychological factors. The idea that society and its rules and institutions have psychological origins in a putative human nature implies that there was ever a beginning of society, born out of pre-social human nature. But this is clearly a myth: 'we have every reason to believe that man or rather his ancestor was social prior to being human'.[44] What people call human nature must therefore be the product of social processes, not the other way around. Thus, sociology is autonomous. The actions of the individuals that make up society can, 'to a very large extent', be explained by what Popper calls the 'logic of the situation' in which they occur.[45] Whatever psychological factors are at play tend to be trivial, and they are never purely psychological.

Characteristic of Popper's intentional disregard of psychology and the personal element is his discussion of leadership. Plato, he wrote, had saddled history with a lasting confusion when he had made the question 'who should rule' the central issue in politics. This is precisely the wrong question: we shouldn't ask who is a good ruler; we should rather ask how we can get rid of bad rulers. Even if we should decide that only the best or the wisest or the bravest should rule, we have to keep in mind that sometimes people do not live up to our expectations, that even the best can be corrupted, and then we will want a way to depose them. '[W]e should always prepare for the worst leaders, although we should try, of course, to get the best.'[46] Checks and balances are the essence of democracy, not a particular kind of leadership. We should look for an institutional answer, not a 'personalist solution', to the question of politics.[47]

What's more, the idea that one can educate or select the best leaders is self-contradictory. Intellectual excellence lies in the spirit of criticism, in originality and independence, and is therefore at odds with 'any kind of authoritarianism'.[48] Authorities will always select the mediocre because they will never admit that the ones who rebel, who resist their influence, are the truly excellent. We must not burden educational institutions with the task of educating and selecting leaders, whether in science or in politics. That will turn education into a hurdle race in which students study for the sake of passing exams rather than for the sake of learning per se. It is this impossible goal of selecting the best that has made our secondary schools and universities such a 'devastating system of education'.[49]

Mario Bunge has noted that Popper's institutional approach to democracy does nothing to help voters, who are precisely faced with the question that Popper dismissed as misleading – who should rule?[50] Equally, his critique of the educational system and its disastrous preoccupation with testing and selection leaves open the question of whether leadership can be taught and how. More generally, because of the focus on institutions and the total exclusion of the 'personal element' from the equation, social engineering remains a bloodless, largely abstract affair of piecemeal trial and error in Popper's social philosophy.

It is, or should be, a rational process of scientific experimentation and critical reflection. But as we shall see in the next chapter, while Popper formulated this institutional approach to social technology and engineering, elsewhere psychologists addressed precisely the questions that Popper left unanswered. How should one deal with the human factor in social engineering? And to whom should we trust this task – who is a good social engineer? Leadership was a central concern for these psychologists. How can one lead democratically, and how must democratic leaders be trained?

7 Tactful Leadership

For Popper, as we have seen, human engineering was anathema. Social technology is a technology of social institutions, not of human behaviour or human relations. Ameliorating society is a matter of improving institutions, the quasi-machines that make it up. Their functioning must be improved through piecemeal structural reforms, but the people who work in those institutions or who depend on them must be left alone. The personal element may be frustratingly irrational and unpredictable, but to tamper with it is not only bound to fail, worse it is undemocratic and unscientific. But while Popper was developing these ideas in New Zealand, precisely such a psychological version of social technology was taking shape in the United States, in the group of social psychologists around another émigré: the German Kurt Lewin.[1] Although their social technology did target the personal element, for Lewin no less than for Popper the aim was to join the interests of science, technology and democracy.

Socialising Taylorism

Born in 1890, Lewin was twelve years older than Popper. He studied at the University of Berlin under the experimental psychologist Carl Stumpf. After serving in the army in the Great War, Lewin returned to Berlin's Psychological Institute, where he came under the influence of the Gestalt psychology of Wolfgang Köhler, Kurt Koffka and Max Wertheimer. Although he never became an orthodox Gestaltist, their holism in particular greatly influenced Lewin's approach to social psychology. Lewin's interest in social psychology was first awakened by his studies in psychotechnics, beginning in the war. One of his first articles was a defence of psychotechnics against the charges made against Taylorism (see Chapter 3). As it was in the United States, scientific management was a topic of intense debate in Germany.[2] In *Die Sozialisierung des Taylorsystems* Lewin notes that 'not only Taylorism, but any kind of psychologisation' meets with suspicion from socialists because it appears to treat human beings as mere machines and numbers in a calculation.[3] Psychotechnics seems to serve the goal of capitalist exploitation. This, however, is not the fault of the psychological approach, Lewin insists: psychotechnics is merely an instrument that can serve the ideal of a socialist, egalitarian society just as well. From a socialist standpoint, moreover, the worker is a means of production no less than the machine he works with, and as such his talents and labour belong to the community. Socialism demands that the worker's capacities are put to the best possible use in the service of the common good. Psychotechnical techniques of vocational guidance, selection and training have a vital function in a socialist society.

What distinguishes a socialist use of psychology from Taylorism, Lewin writes, is that it recognises that the worker is not just a producer but also a consumer of labour. Work is not only and not always a burden to him but also a source of pleasure and fulfilment. In a socialist society, these two aspects – labour as production and as consumption – must be in balance. One way to guarantee this is to involve the workers in the psychological study of their work. Psychotechnical studies are best conducted on location, in the factory or workshop concerned, in collaboration with the management and the workers themselves. Just as in laboratory experiments, the subjects in these studies, the workers, have a role as 'Selbstbeobachter' (self-observers) and supply introspective reports of the effects of the experimental manipulation of the production process. There is nothing degrading in being an experimental subject; on the contrary, it can be very instructive: 'People who have served as subjects in experiments will confirm that it is, as a rule, quite an interesting experience and will often teach you something about your personal way of working as well.'[4] Lewin never referred to this essay again,[5] but the idea of democratising science and technology by directly involving the people who were affected would return in his later work.

In the 1930s Lewin developed a Gestalt-inspired approach to the analysis of the behaviour of individuals that he dubbed a field theory.[6] The fundamental concept was the 'life space', the total constellation of psychological forces influencing an individual at a given time. In the mathematical terms that Lewin liked to use: $B = f(LS)$, or $B = f(P,E)$. Behaviour is a function of the life space, and the life space in turn is a function of the relation between person and environment. Behaviour is caused not by the objective situation but by the individual's perception of it as well as by his beliefs, needs, desires, goals and so on. The life space is a dynamic psychological force field, and if one knows this force field in sufficient detail, one can predict what the individual will do next. Lewin developed a system for representing the structure of the life space based, albeit rather loosely, on mathematical topology. Life spaces were represented as egg-shaped figures, subdivided into regions representing the various forces that attracted and repelled the person. Later, when Lewin turned his attention increasingly to groups, they too were seen as dynamic wholes made up of interconnected parts. A group is not just a collection of individuals; it has a reality of its own. It too has a life space: a dynamic force field resulting from its goals, its values, and the way it perceives its situation. Social change is possible by moving groups from one 'quasi-stationary equilibrium' to another.[7]

In a discipline that was, certainly in the United States, increasingly dominated by the associationism of the behaviourists, Lewin's ideas were highly unusual, but nevertheless Lewin's fame as a creative experimentalist and original thinker began to spread beyond Berlin. In 1932, at the invitation of Lewis Terman, Lewin spent six months as a visiting professor at Stanford University in California. Travelling back to Germany via the Soviet Union in January 1933, he received news that Hitler had become chancellor. Lewin was Jewish, and unlike Popper he was also a Zionist. He had been following the rise of Nazism during the 1920s and early 1930s with increasing worry. Now that Hitler was chancellor, Lewin immediately decided to leave Germany for the USA, as did Köhler, Koffka and Wertheimer. After two years at Cornell University, Lewin secured a position (financed by the Laura Spellman Rockefeller Fund) at the Iowa Child Welfare Research Station, a research institute of the University of Iowa.[8]

Democratic Leaders

With the developments in Germany in mind and inspired by his encounter with American society, Lewin developed an intense preoccupation with democracy as a scientific problem. Arguably the most famous product of this is a set of experiments that he conducted with his student Ronald Lippitt in the late 1930s.[9] Lewin and Lippitt set out to answer the question of to what extent the behaviour of a leader influences the group. Small groups of boys were given the task of making theatrical masks under the direction of either a 'democratic' or an 'authoritarian' leader. The most significant event during the first experiment was the accidental production of a third leadership style, the 'laissez-faire'. Ralph White, another graduate student, was at one point charged with playing the role of democratic leader but misunderstood it to mean an aloof, laissez-faire stance. To Lewin's delight, White's group quickly descended into chaos. Laissez-faire leadership was added as a third condition for a follow-up experiment, and it reliably produced mayhem and aggression among the boys. Lewin concluded that true democracy requires guidance. It needs to be managed at the level of the group. Thus, what to Popper was anathema – the moulding of individual behaviour and personal relations – became the focus of Lewin's research programme, but the goal was the same: to secure and promote democracy with the help of science.

Lewin's programme had an explicitly technological character: it aimed to solve practical problems as well as develop and test theories, the studies were often conducted in the setting where the problems had arisen, and questions of technique and its transmission through training took centre stage. Over time, engineering terms started to enter the vocabulary with which this technology was discussed. At the same time, the techniques that were developed were tactful: they were often ways of guiding individuals and groups without giving offence or incurring resistance. They were subtle and indirect techniques, and effective *because* of their subtlety. These two aspects of the work of Lewin's group are closely related to another pair: formality and sensitivity. There was in this research programme on the one hand a strong drive towards formalising theory and practice in the form of diagrams, mathematical models and lists of principles and instructions. On the other hand the experience of the individual in the group was always the primary consideration, and developing the sensitivity of both group members and trainers was an important aim. Historian William Graebner has criticised Lewin for turning democracy into an instrument of social control and for making that control invisible.[10] What masquerades as democracy is actually coercive because it is 'democratic in method rather than in purpose'.[11] Elsewhere in the same article, however, Graebner writes that 'democratic social engineering was neither clearly coercive nor clearly noncoercive'.[12] I think tact and technology are a better pair of concepts for describing these practices than coercion versus democracy.

We can see the interplay between these two aspects, tact and technology, most clearly in the work of Lewin's graduate students, the people most closely involved with the research subjects. One student in particular, Alex Bavelas, was an expert in dealing with people, and as such he was a major influence on Lewin's research programme. Judging from descriptions of him, Bavelas appears to have been something of an intuitive master of tact. According to Dorwin Cartwright, 'he was just a genius at doing group work'.[13] Alvin Zander has described Bavelas as a 'charmer', who got better results from his experiments than other students because he got his subjects much more involved.[14] Lewin greatly admired the way he handled people and used him as an example for other students. 'Lewin would have Bavelas lead groups – children's groups – behind oneway glass so we could watch him. Lewin would stand there and watch him and get all excited and say, "See what he is doing, see what he is doing." Then Bavelas would come out and Lewin would tell him what he had been doing and Bavelas would say, "I didn't know that."'[15]

Training Democratic Leaders

Bavelas joined the Iowa Child Welfare Station as a graduate student. Discussing subjects for his MA thesis[16] with Lewin, Bavelas first proposed a number of projects that would build on the work of earlier students of Lewin like Bluma Zeigarnik to fill in the gaps and work out the details. These ideas met with little enthusiasm. Lewin considered such work to be 'like nibbling away at an infinite cake'.[17] However, Lewin's interest was suddenly piqued when one day Bavelas remarked, a propos the Lewin and Lippitt studies of leadership styles, that he thought it should be possible to train people to adopt one or the other style, 'that you could change their behavior deliberately'.[18] When Lewin put him on the spot and prompted him to explain how he would do it, Bavelas hesitantly sketched out a few possibilities, and Lewin simply said, '[F]ine, we'll do that'.[19] Lewin was always willing to take up even a half-baked idea and develop it further as long as it seemed conceptually interesting.

The experiment took place in the Jewish Community Center in Des Moines during its summer 'Home Camp', where children were offered recreational activities – clay modelling, making dolls, playground games, etcetera – during the school holiday. The various activities were supervised by leaders supplied by the Work Projects Administration (WPA), a New Deal agency that put to work unemployed people in public works. The question that Bavelas and Lewin set out to answer was whether mediocre leaders can be retrained into 'efficient democratic leaders'.[20] Six definitely unsatisfactory leaders were picked by their WPA supervisors and divided into two groups of three, an experimental and a control group. All leaders were then tested by observing and recording their behaviour with the children, specifically how they 'controlled the children': whether they gave commands ('direct control'), directed the group by praise or directed by giving the children the responsibility of making their own choices.[21] This last 'initiative-stimulating method' was used only rarely. This was the democratic leadership that the training programme aimed to instil.[22]

In a brief report of the experiment that Bavelas wrote together with Lewin, they say very little about the actual training, but a longer report by Bavelas alone gives a rich view of his subtle approach to training the leaders. Bavelas first of all describes the three trainees and their background and personality and notes the situation that they and their colleagues are in: working for the WPA is a low-status job with little security.[23] They could be fired at any time with only two weeks' notice. As a consequence, they are very suspicious about the training at first, feeling observed and examined, particularly by the two stenographers present at the meetings to record the process. In the fourth meeting, a change in the trainee group becomes apparent: they start to pour their hearts out about the centre and about other leaders and begin to act and feel like a group. By the end of the experiment, they are enthusiastically making plans for the future – a better job, more education – and all would like the training to continue. By that time, the groups of children they lead are loyal to them, cooperative, productive and full of initiative. The non-trainee leaders in the control group, on the other hand, have become jealous and even aggressive towards the trainees.

This transformation is wrought by a training method that exemplifies the kind of leadership that it aims to impart. 'It was the intention of the present experiment that the trainees should be able to make use in their later work of their personal experiences as members of a democratic group during training. The experimenter, therefore, was in the role of group leader; in fact, his conduct had to be illustrative of the type of leadership the trainees were learning to use.'[24] (Bavelas speaks of himself in the third person in the paper.) Rather than rigidly following a programme, Bavelas had a number of points for each meeting that shouldn't be overlooked, but allowed the group to make its own way through the problems and develop the techniques to solve them itself. At first, however, Bavelas avoids the issue how to handle particular situations, instead encouraging discussion about the principles underlying interventions. 'The result was that techniques which the leaders actually used later in working with groups never became a "set of tricks", but, rather, remained flexible and easily modifiable to meet specific situations.'[25] Bavelas knew the importance of *kairos*: the actions of a leader must always be adequate to the particular people and situation he is dealing with.

Having established the primacy of 'attitudes' over 'techniques', Bavelas proceeds to 'sensitize' the trainees to the concept of leadership itself by asking them to make a list of qualities of a good leader.[26] Rather than criticising the vagueness of the terms they come up with (such as 'tolerance' and 'patience'), he asks them to use the list to describe another leader, which results in widely differing assessments. After further discussion a new list is drawn up, tried out and so on for several more iterations until a sufficiently precise and 'realistic' list ensues.[27] Next, the group is asked to draw up a list of objectives of its work. Bavelas again refrains from evaluating the proposal – 'the production of good craft articles, the entertainment of the children, and the keeping of good discipline' – and he notes the group members' ideas about how to attain those objectives without further comment.[28] Instead he asks them to consider other possible goals, which results in a long list of objectives of varying generality. The group then starts to structure this collection of goals, working out how they are related, which are primary and which subordinate, which objectives (an atmosphere of enthusiasm, say) also bring about others (discipline and entertainment). Bavelas then allows the discussion to return to 'a study of the "paths to the goals" – in other words, techniques'.[29] Bavelas's main procedure for helping the group develop techniques is observation followed by discussion and evaluation: group members watch him at work leading a group, watch each other, watch films of the Lewin and Lippitt experiments and are made to experience different leadership styles when Bavelas, without warning, changes the way he leads – all followed by discussion and criticism of the techniques they had observed. Each day ends with a thirty-minute clinic for further discussion.

What stands out in this report is, first of all, how unobtrusively Bavelas directs the group members to where he wants them to go. He never tells them what to do or think but changes the direction of the discussion at crucial points or introduces a new question or topic. He lets the group discover the facts by itself but does make sure that it looks in the right direction. Democratic leadership, the kind that Bavelas practices and wants to 'induce' in his pupils, is paradoxical: it aims to erase itself. Democratic leaders get their subjects to make their own decisions and take responsibility for their own choices. The tactics and techniques involved must be subtle. To the extent that democratic leadership is directive, it is so in an indirect way. To borrow Javier Lezaun's formulation, democratic leadership must be forceful yet imperceptible. This requirement gives Bavelas's own leadership its ghostly character.

Secondly, it is remarkable how steadfastly Bavelas guides the group's attention away from concrete ways of handling situations to higher-order considerations: to principles, attitudes, leadership style and objectives. The question of how one should lead a group of children in a day camp – keep them cheerful but disciplined, make them work together without conflicts, solve the conflicts that inevitably arise – is only considered in the latter stages of the training. First the group needs to learn what the principles behind the techniques are, what goals they serve, what style of leadership they express. At the end of the report, the reader still hasn't learned what techniques the group eventually develops out of its observations and discussions. Likewise, few details are given about Bavelas's own actions. Bavelas describes his approach in the first place at a strategic level. The core of his strategy is that he will practice what he preaches; his own style of leadership must exemplify the one that the students must adopt. He therefore doesn't lecture but makes the group discover the facts that he wants it to know. This strategy is realised in a number of techniques, such as the observe-and-discuss routine and the propose, try out and refine cycle. Both are procedures that facilitate the group's own journey to knowledge. Bavelas's finer-grained manoeuvres, however, are indicated with rather vague terms. Bavelas speaks for example of 'eliciting' ideas from the group about how to attain their objectives and 'inducing' the trainees to think of other possible objectives.[30] Not only do these terms indicate indirect, subtle forms of interaction compared to, for example, 'asking', 'demanding' or 'telling to', but what they consist in, moreover, is not specified. How one goes about eliciting and inducing is not made clear: no script, no choreography is given.

The Group Decision Method

One of the most important techniques of democratic leadership to be developed by Lewin and his students was the group decision method. It originated in a series of studies commissioned by the Committee on Food Habits of the National Research Council. The United States' entry into the war had necessitated food rationing, and in the interest of good nutrition, it was important that people were willing to eat the foods that were available. This was seen as a problem that required the expertise of social scientists, and the anthropologist Margaret Mead was asked to serve as secretary to the committee.[31] Through Mead, her close friend Lewin was tasked with designing a series of studies on how to change food habits. As was always the case in Lewin's work, the research was a group effort, the fruit of the informal, lively and creative discussions that Lewin liked to have with his students and staff.

The studies began with a preliminary investigation of the actual food habits of Americans. A group test was devised that employed the same techniques that Bavelas had developed for an earlier study of the 'group ideology' of schoolchildren.[32] Instead of asking direct questions – what do you think is good behaviour, what foods do you like – which might yield socially desirable answers, the psychologists asks what a member of the respondent's group would typically think or like. The assumption was that the answer would reflect the individual's own opinion. 'It seems that the indirect method actually solicits more open and personal answers by setting up a situation which from the subject's point of view is impersonal.'[33] The indirect method circumvents possible resistance of the subject, who might fear criticism if she condones vandalism or professes a liking for unhealthy food.

Graebner has traced the use of group processes as a vehicle for what he calls democratic social engineering – 'a new form of authority distinctly nondirective, participatory, persuasive' – back to the end of the nineteenth century.[34] In so-called foremen's clubs, for example, superintendents used carefully guided group discussion as a way of maintaining foremen's allegiance to the company rather than the union but also to train them in the use of the same democratic leadership when dealing with workers. Working with Lewin on the food studies, Alex Bavelas came up with a crucial innovation, 'the cluster of ideas and procedures that became known as "group decision"'.[35] In the preliminary report of the study in which this technique was first used, Lewin explains that knowledge alone does not suffice to change behaviour. That requires motivation as well, but motivation in turn has to be 'brought down from the level of wishes and sentiments to the level of a "decision"'.[36] The experiment therefore pitted a half-hour lecture by a nutritionist against a group discussion plus decision led by Bavelas and measured how many of the housewives in the group had actually prepared a novel food in the week after the meeting. To put the method to the severest possible test the foods in question were hearts, kidneys, and brains, which were almost universally disliked for reasons of 'physical aversion, social status, and superstition' (in other words, eating brains, etc., is gross, lower class and spooky).[37]

Lewin describes the group discussion/decision procedure as one in which the group is gradually led, 'through a free group discussion', to acknowledge the need to change food habits and to accept responsibility for doing something about it.[38] To see how Bavelas solved the tension inherent in such a process ('leading' through 'free' discussion), we must turn to the detailed description of the method in appendix II of the report. Noteworthy is first of all again the use of the indirect method. After 'creating a rather sober atmosphere' and linking the need to change food habits in times of war to patriotism and good citizenship, Bavelas tells the group about a method for changing people's habits – a 'face-to-face approach to groups of housewives like themselves' – and then poses the question 'do you think they would cooperate?'[39] The whole group answers in the affirmative: women like themselves would cooperate.

Having a foot in the door, Bavelas reveals the foods in question, and quickly, so as not to give the women opportunity to express their disgust, takes a census: who has served these meats before? Few have, of course, and to defend what they now understand to be a lack of culinary patriotism, the group members spontaneously begin to offer reasons why they haven't. At this point the nutritionist is brought in to briefly explain how to prepare the foods so as to make them less offensive. Mimeographed recipes are handed out. Then Bavelas asks the group members to indicate by a show of hands whether they will try serving these foods, taking care to ask the question for each specific meat in turn. Having already said that women 'like themselves' would cooperate and thoroughly imbued with the national importance of the matter, most women publicly commit themselves to serving their family at least one of the dishes. Thus, there is no 'group decision' as such; rather, the individuals decide in a group setting, which 'gives the incentive for the decision, and facilitates and reinforces it'.[40] Finally, to make them feel even more responsible for their decision, Bavelas suggests the group works out a way of letting the nutritionist know how successful they have been. Dutifully the group begins to arrange a follow-up meeting.

As in the leadership training experiment, the group process is free only within strict limits set and guarded by Bavelas. In fact, the group is allowed to discuss only two points: concrete reasons for not serving offal and 'methods of presenting food to the family' that have proven successful in the past and might be tried with these meats.[41] The discussion about 'reasons' is crucial because it forces the group to articulate specific objections to offal, which can then be neutralised by the nutritionist with the appropriate culinary tricks, leaving the group no further arguments for not trying the meats. The discussion about ways of presenting foods merely serves to reinforce the group members' decision to try them. Apart from these brief spells of 'free' discussion, all the steps in the procedure are taken by Bavelas. Nor is the decision finally taken by the housewives entirely free: 'Following the show of hands, which may be few or many, the leader says again that it doesn't matter which they try so long as they would agree to try one.'[42] 'No attempt is made to manipulate the group by high-pressure sales talks,' writes Lewin,[43] but it seems clear that the emphasis in that statement should be on 'high-pressure': Bavelas overcomes the resistance of the housewives with subtle rhetorical manoeuvres rather than sheer discursive force, but the group is being led all the same. In his concluding remark, Lewin makes clear that democracy in this study is an engineering instrument, not a goal in itself: 'Methods similar to this may prove to be an efficient means for democratic action in various fields and may provide a better channel for the much-needed translation of expert knowledge into social action.'[44] It should be particularly effective, he adds, if one applies it to several strategically chosen groups in the community at the same time.

Discovering the Facts

In 1939, Lewin had been invited by his friend and admirer Alfred J. Marrow to visit a garment factory in Marion, Virginia.[45] The plant was the latest extension of the Marrow family business, the Harwood Manufacturing Corporation. In it, some 300 inexperienced young women from rural Virginia made pyjamas, but they struggled to meet the production targets, and many left even before they finished their training. Marrow, himself an industrial psychologist, wanted Lewin's advice about the problems at the plant.[46] During his three-day visit, Lewin came to the conclusion that the problem lay not so much with the workers themselves but with the way they were managed. He recommended changes in the behaviour of the supervisors and in the way the production targets were set. Later, when these changes had begun to lead to higher production, Lewin was able to persuade Marrow to start a programme of research on 'human factors in factory management'.[47] The 'Harwood Studies', as they are called, continued into the 1970s.[48]

Alex Bavelas was the first of Lewin's graduate students to be sent to Marion. During his time at the plant, Bavelas further developed the group decision method and that of leading a group to 'discover the facts for itself', which we saw him use in the leader training. Bavelas's handling of both techniques is described in loving detail in what must be the oddest text in Kurt Lewin's oeuvre, a short paper published in the *Proceedings of the Second Brief Psychotherapy Council* titled 'The Solution of a Chronic Conflict in a Factory'.[49] That Lewin devoted an entire paper to a single event – Bavelas solves a conflict that erupts one day in the Harwood plant – is testimony not only to his ability to see theoretical grandeur in small matters but perhaps also to his admiration for the skills of his student. Lewin's text stands as a monument to Bavelas's tact and techniques.

'The Solution of a Chronic Conflict in a Factory' is a text with many layers. At the bottom are the events that had taken place that day early in the afternoon and that had led to a quarrel. This incident is only known from the reports of the two quarrelling characters, and these reports in turn are passed on by Alex Bavelas, who also narrates how he defused the quarrel and solved what according to him was the underlying issue. Finally, Lewin presents Bavelas's report and interprets the conflict and its solution as an illustration of 'group dynamics', a theoretical concept he had been developing since 1939. Thus, in the end it is Lewin (the sole author of the article) who gets to say what actually happened. In fact, in his interpretation the original incident and the person who set the whole thing in motion become irrelevant.

Bavelas's report is written in the form of a play, with acts, scenes and a cast of characters but, apart from a few quoted statements, without dialogue. The scene of the incident is described as a sewing factory employing about 170 'girls' to operate the sewing machines, five 'floor girls', one mechanic named Paulson and one supervisor named Sulinda (her first name) with a boss, Alanby, at the top of the hierarchy.[50] One afternoon, at 1:30 PM, Alanby calls Bavelas into his office, where he finds Paulson and Sulinda, both looking uncomfortable. Alanby explains that Paulson and Sulinda disagreed on which machines had to be repaired first and that one of the girls was gossiping and setting them against each other. Paulson and Sulinda both seem on the verge of tears. Bavelas tries to defuse the tension by making a 'trifling joking remark' and lights a cigarette 'to gain a little time'.[51] He offers cigarettes, but only Alanby takes one. He sits down on the corner of the desk, but 'this did not noticeably break the rigid atmosphere'.[52] Having failed to lighten the mood, he asks Alanby what's going on, but when Alanby's explanation brings Paulson and Sulinda close to tears, he quickly interjects his expert opinion – such behaviour in operators is quite common, he says, and he mentions an occurrence in a factory where he had worked earlier. Then he looks at his watch and says he shortly has a brief appointment with an operator but that afterwards he would like to discuss the problem in detail with each of them.

The appointment is a ruse: after leaving the office and briefly talking with Sulinda while walking upstairs, he returns to the office to ask Alanby about his view on the matter. Alanby tells him that quarrels like this happen occasionally between the two because Paulson and Sulinda have incompatible characters, but this time they both threatened to walk out, and he asks Bavelas to 'smooth things over'.[53] Bavelas then starts his investigation. In his interview with Sulinda, she begins by criticising Paulson's capabilities as a mechanic, but then she recounts what had happened earlier that day. One of the operators had come to her saying that Paulson refused to repair her machine. When Sulinda confronted him about this and told him he had to do it, he got angry and said it was a lie that he had refused to repair the machine. He then went to the girl in question, who in turn told him that she had never said that and Sulinda was lying. Next, the two of them had come back to Sulinda and had accused her of lying. At that point, Sulinda got her coat and went to tell the boss that she was quitting.

Although 'the lie', as it is called in the rest of the paper, is obviously the heart of the matter for Sulinda, Bavelas shifts the focus by asking 'factual questions about the frequency of breakdowns and whether they were more frequent in certain types of machines, etc.'.[54] It quickly becomes clear that Paulson, the sole mechanic in the factory, has far too much work on his hands, leading to irritations between him and the operators, who don't make any money when their machines are broken. Next, in his interview with Paulson, Bavelas has no trouble getting him to agree on this diagnosis of the underlying problem. Bavelas proposes to Sulinda and Paulson that he will talk with the girls who complain most to hear what they think about this problem. Both agree this is a good idea.

Next (act III, scene 3), Bavelas has a brief interview with the girls in question. Bavelas gives few details about this part of the events, aside from reporting that they agreed with his diagnosis and the solution he proposed: Paulson is too busy, and the girls who are having the most trouble should get together with Bavelas to work out a way to reduce the time they have to wait for repairs. So it is done, and in the group meeting Bavelas presents the problem, which is now rephrased as 'that at certain times when more than one machine broke down there was a shortage of mechanic services',[55] and the constraints on a solution, namely that 'it was unlikely that another mechanic would be hired in view of the difficulty of deferring even Paulson from military service'.[56] Bavelas then proceeds to 'stimulate group discussion' on how to allocate Paulson's services when more than one machine breaks down at the same time.[57] The group works out a decision algorithm (first come, first serve, unless one machine is more 'critical' than the other), which is then presented first to Paulson and then to Sulinda. Bavelas suggests that it should be Sulinda who takes the crucial decision of which machine is more 'critical'. Both accept the solution but express doubts that the other will be happy. Next, Bavelas goes through the new procedure in detail with the entire group of operators, and then it is implemented. The play ends with a brief epilogue set a few weeks later, in which Alanby tells Bavelas that relations between Sulinda, Paulson and the girls were much better and that Paulson seemed to have more time and had 'installed a system of loudspeakers in the factory and played recorded music two periods a day', which had improved the atmosphere even further.[58]

Lewin's commentary is interspersed between the acts. He explains that the lie was such an important issue for Paulson and Sulinda because it concerned their authority with the girls and Paulson's position of equality with Sulinda. He praises Bavelas's deft handling of the power relations, like when he asked Sulinda's consent before he talked with the girls. But the key to defusing the situation was the way Bavelas had shifted the focus of the interviews with the mechanic and the supervisor from the lie to the issue of Paulson's workload: 'By reorienting Sulinda's and Paulson's perception from the field of personal emotional relationship to the same field of "objective" facts, the life-spaces which guide the action of these persons have become more similar although the persons themselves are not yet aware of this similarity.'[59] Focusing the two on something they could agree on – Paulson was too busy – had revealed the overlap in their life spaces, and now both could become part of the same plan.

Next, Bavelas had refocused the girls as well. They too were turned 'toward the objective difficulties of production'.[60] Their support, Lewin explains, was essential. They do the work and know best what the problems are, but because they are lowest in the hierarchy, any rule or view suggested from above will be seen as an imposition. The solution had to emerge from the group itself. Bavelas's decision to pick what Lewin calls 'the trouble-makers' (note the shift from having trouble to making trouble) for the group discussion was wise because although they are probably less objective, they are more important 'for the group dynamics in the factory', and, being troublemakers, they would resist any solution that they hadn't come up with themselves.[61] Instead, '[s]ince the rules are developed by the girls themselves, their acceptance is implied and strongly entrenched'.[62] The way that Bavelas gets Paulson and Sulinda to agree with the plan also betrays his insight into their life spaces: 'The reality is presented correctly, but those aspects are brought into the fore which are linked with the psychological situation of the person in question and are helpful in bringing about favorable permanent motivation.'[63] For example, he doesn't speak to Paulson of dividing authority but of freeing him from the burden of having to make decisions amid the clamour of a group of girls, each wanting their own machine fixed – 'a decision situation providing the possibility of high emotional tension', as Lewin puts it.[64] At the same time, Sulinda will not be giving him orders: she only decides which machine is more critical. What Paulson should do with that information follows from the rules agreed upon by everyone.

Lewin sums up the idea behind Bavelas's intervention as '[t]he realistic demands of production have to be satisfied in a way which conforms with the nature of group dynamics'.[65] It is not enough to determine a problem in the production process (in this case, the bottleneck in machine repair) and design an efficient solution for it. These facts, objective as they may be, have to be accepted by the people involved. The actions of an individual are not determined by the facts but by his perception of the facts. 'It is correct that a "sufficiently objective" picture of the production channels and problems should arise from the investigation. But it is equally essential to realize that the "subjective" view of the participants counts most.'[66] The best way to achieve this alignment of objective facts and subjective perception is to let the group do the fact-finding itself. The machine operators are made 'the fact-finding body'.[67]

Lewin notes that, as the afternoon wore on, the lie had 'evaporated into thin air without ever having been treated directly'.[68] By shifting the perception of the situation to the problems in the production process, the lie had 'lost the character of a "fact"'.[69] Lewin's metaphor is an apt description of the events of the afternoon. It seems fair to say that the whole saga is one of smoke and mirrors, with Bavelas in the end as the supreme magician. What started with an operator playing tricks on Paulson and Sulinda turns out in fact to be a production issue. Paulson, Sulinda and the girls believe they discover this fact on their own, but it is Bavelas who leads them there. Next, the troublemakers, stimulated by Bavelas, devise the solution (first come, first serve, unless one machine is more 'critical' than the other), but Bavelas later slips in the crucial component: who gets to say which machine is 'critical'. From the moment he first distracts Paulson and Sulinda with his fake 'appointment', Bavelas unobtrusively orchestrates each event. Along the way, the lie evaporates, but no one notices, everyone being enthralled by the facts and solutions they appear to find on their own. Bavelas's greatest trick is to make his own contributions to the process invisible.

We have only Lewin's and Bavelas's word for the events that afternoon between 1:30 and 5:00 PM. Perhaps what seems like a brilliant feat of magic from Bavelas was actually a more complex process in which he was being played by the girls as much as he played them. Or perhaps Alanby simply used Bavelas to solve a personnel issue that he didn't want to burn his fingers on himself. Without other sources for the events it is impossible to tell. Based on this text, in any case, we can interpret Bavelas's magic as an example of tact as defined in Chapter 1 of this book. Bavelas is a human engineer who employs a strategy, which includes deception, to prevent or overcome resistance to the changes he wants to introduce. The main obstacle he has to contend with is the obligation to be free typical of modern individuality: he cannot simply tell people what the facts are and what has to happen. Changes will only stick if people consider them their own project.

Theory and Technique

Among Lewin's more enduring contributions to social psychology is the motto 'there is nothing so practical as a good theory'. As Michael Billig has shown, the line has become part of the discipline's common sense, underscoring its claim to practical relevance.[70] It is doubtful whether Lewin would have been impressed with the kind of theory that social psychology produces nowadays – he was averse to the use of statistics to describe quantitative differences between individuals or groups and considered his field theory to be scientifically more advanced. In any case, Lewin's use of the phrase (which in fact can be traced back to Kant)[71] and its continuing popularity among social psychologists raise the question what relation theory and practice actually have in social psychology.

Lewin translates Bavelas's actions into the concepts[72] of field theory, but could the process be reversed, and the theory used to specify how the psychologist should act in a situation like the conflict between Paulson and Sulinda?

Lewin raises the question himself when he notes that 'it would be interesting to discuss in detail to what degree Mr. Bavelas followed a preconceived plan in his treatment and to what degree he let himself be guided by circumstances'.[73] Without actually going into detail, Lewin writes that Bavelas had started by cautiously feeling his way, but once the group discussion with the operator had produced the new rules, he had 'considered the rest of the procedure routine, the step of which he could quite well see in advance'.[74] Bavelas had started by improvising and finished with a routine set of steps.

When we look over the reports of Bavelas's work during his time with Lewin, we see that the extent to which his interventions were planned varied, as it did here in the space of one afternoon, but also that his actions were seldom scripted or choreographed. In 'A Method for Investigating Individual and Group Ideology' Bavelas gives precise instructions for the phrasing of the questions that constitute the method, but it is an exception.[75] Usually the actions are prescribed at a lower resolution, as they were in the leadership training study, where the focus was on objectives, attitudes, principles and style. In the food habits study, the procedure is described in two ways: one column of the text gives a more or less concrete description of 'what takes place' with next to it 'the meaning of this step'.[76] Moreover, 'what takes place' is described as a series of actions ('The group leader emphasizes the fact that the problem of nutrition has been found to be acute'), not behaviours and utterances.[77] It is left to the group leader to put that emphasis into words.

The techniques of group dynamics are general principles, which the technician (a group leader, for instance) must actualise in a specific way in each specific situation. Theory informs practice only up to a point or, rather, down to a certain level of resolution. The technical details must be filled in by the moderator, trainer or leader. There is no script and no choreography. Any specification of a technique is always conditional; the principles, attitudes and so on remain the same, but their realisation may vary with the particular situation at hand. Thus, the technician is an inherent part of the technique. Lecturing and rote learning are of limited use in the training of these techniques because there are no specifics to lecture about or to memorise. In the leadership training study, once the principles had been impressed on the trainees, the rest of the training consisted of observation followed by discussion.[78] Theory and practice do not meet in the form of a list of specific, theory-based instructions for action. Rather they come together in a certain sensitivity in the technician, what one could call their sense of *kairos*. Such a form of technique cannot exist in the form of prescriptions; it must be developed through observation, discussion and the enacting of exemplary situations and interventions and become embodied.

Observation and discussion were already a central training method in Lewin and Lippitt's experiments. As Lezaun and Calvillo noted, to calibrate the performance of the four leadership styles 'required much more than a mere list of instruction about what the different "leaders" should or should not say. It implied the development of a set of techniques for the embodiment of different sorts of political authority, something that could only be accomplished through an intense process of self- and other-observation.'[79] The result of this training in the Lewin and Lippitt experiments was that the leaders exhibited similar gestures and mannerisms: the film recordings that were made of the experiments show each authoritarian leader dressed in suit and tie, and employing brusque directive movements, for instance.[80] Such standardisation was required to control for the influence of the various personalities of the experimenters, but it was also made possible by the fact that each experimenter could operate in roughly the same situation: the same experimental setting (an attic of one of the University of Iowa's buildings) with similar subjects (twenty young boys of a school affiliated with the university).

However, when the situation is not, or less, standardised, as in the case of the leaders that Alex Bavelas trained, such a 'radical stylization' of a leader's behaviour is not possible.[81] The less standardised the situation is, the less the actions of the leaders can be specified, and the more the leaders have to translate theory into practice themselves based on their sense of *kairos*. What Lewin and Lippitt's leadership experiments produced, conclude Lezaun and Calvillo, was a way of transporting democratic leadership into contexts outside the laboratory: into management training, psychotherapeutic communities and social work. What Bavelas's work suggested, I would add, is that this required more than the embodiment of democratic leadership in a set of gestures and mannerisms. What had to be embodied in the leader was primarily an attitude or sensitivity that allowed her to interpret a diversity of situations and translate the principles of group dynamics into actions appropriate to that particular situation at that particular time.

The techniques of group dynamics that were developed in the early 1940s, I suggest, owed much to the combination of Kurt Lewin's theoretical acumen and Alex Bavelas's natural talent for group work. According to the industrial psychologist Mason Haire, '[t]he distinctive contribution [Bavelas] made was this on-the-spot sensitivity to what was there . . . and the artistic sensing of the phenomenon. And it was Lewin who formalized that into a theoretical thrust. But I think it was just a question of having somebody who could do it well enough so that it could be isolated and dealt with.'[82] A feeling for *kairos* was required to get the development of these techniques off the ground as much as it remained needed in their application. Ironically, after finishing his work with the Harwood Company, Bavelas went to the Massachusetts Institute of Technology in 1944, shifted his focus to laboratory studies and got his PhD for a thesis on 'Some Mathematical Properties of Psychological Space', in which he presented an analysis of life spaces that was mathematically much more rigorous than Lewin's.[83] He was an early convert to information theory and cybernetics.[84] He is now best known for a study of communication patterns in task-oriented groups, in which he used laboratory experiments guided by geometrical models to compare the efficiency and effectiveness of groups with different structures of information exchange.[85]

Engineering

Lewin arrived at MIT a year after Bavelas. He had become increasingly unhappy in Iowa.[86] He felt the resentment of colleagues at the Child Welfare Station, who objected to his frequent absence for consulting work and to the fact that as a consequence he neglected such chores as attending faculty meetings. For his part, Lewin came to believe that there was no room at the University of Iowa for the kind of practice-oriented work that he believed was necessary to bring both the field and society forward. After searching his network for interested parties and sources of funding, Lewin eventually moved his group and its activities to two new institutions. The Commission on Community Interrelations, part of the American Jewish Congress, would focus on studies with an immediate practical payoff, but Lewin also wanted room for work that had a theoretical as well as practical aim. This work was housed in the Research Center for Group Dynamics at MIT. Key to convincing the MIT officials that Lewin's group would be a valuable addition were the efforts of Douglas McGregor, professor of management at MIT.[87] McGregor argued that Lewin's work fit the engineering mission of MIT very well, saying that his work in 'social engineering' might well prove to be equally important as recent engineering feats like radar.[88] It also helped that Lewin brought with him a substantial grant from the American Jewish Congress, making the Research Center rather a bargain for MIT.[89]

In a programmatic statement of the Research Center written in 1945, Lewin emphasised the analogy between engineering and the practical work of social scientists.[90] In the statement, Lewin revisits themes he had first worked out twenty-five years earlier in 'Die Sozialisierung des Taylorsystems'. He opens with a quote from Franklin D. Roosevelt: 'the same qualities of heroism and faith and vision that were required to bring the forces of Nature into subjection will be required – in even greater measure – to bring under proper control the forces of modern society'.[91] Lewin calls that control 'social management' and notes that the realisation has only just been dawning that in this field, no less than in engineering, tradition and rules of thumb must make way for 'an understanding on a scientific level'.[92] The social sciences are now beginning to provide that understanding, not just to practitioners but also to the average citizen who is eager to see through the fog of everyday life and find clarification of social issues. But social management is not just analogous to engineering; it is also a necessary complement to it. Engineering has tended to minimise the 'human element' by replacing labour with machines and by treating people *as* machines: 'Where the engineer has studied men he has done so for "selecting" individuals able to handle machines and by viewing the human being itself as a machine. The time-motion study is a marked expression of this aspect.'[93] What is often forgotten is that there is more to running a factory than 'merely setting up production lines'.[94] What has become clear recently is the importance of the group in industry. Group morale and leadership patterns are crucial factors in efficiency that management needs to consider. It is, Lewin concludes, 'entirely appropriate for an engineering school which perceives the main task of engineering in a progressive spirit, to make the scientific study of group life a part of its undertaking'.[95]

Lewin's way of defining and selling the specific contribution of psychology to industry is reminiscent of the psychological reaction to Taylorism thirty years earlier: engineering is a model for applied psychology ('social management'), but at the same time applied psychology is required to correct the deficiencies of engineering, and restore the human element to its proper place in industry. A notable difference with the rhetoric of earlier psychologists, including his younger self, is that selection in the style of Münsterberg and Moller Gilbreth is now also deemed inhuman, and the group gets the emphasis that earlier psychologists placed on the individual. For Lewin, the group is the essence of the human element.

Kurt Lewin died in February 1947 of a heart attack, in all likelihood brought on by his increasingly hectic work schedule. With Lewin gone, MIT quickly lost confidence in the Research Center. Dorwin Cartwright, one of Lewin's former students, succeeded him as leader of the group and arranged a migration of the Research Center (staff as well as students) to the University of Michigan, where it soon merged with Rensis Likert's Institute for Social Research. Under Cartwright's direction, the Research Center continued its mission to reduce the gap between social theory and social practice. Cartwright further developed the technological perspective on group dynamics that Lewin had begun to emphasise around the time of the move to MIT. According to Cartwright, group dynamics was a 'social technology'.

Social Technology

In Cartwright's view, group dynamics revolved around 'change' and 'resistance'. The problems of the twentieth century are problems of human relations, and solving them requires changing people's behaviour. We generally agree on the objectives: peace, freedom, mutual tolerance, etcetera. As Cartwright put it, 'in broad outline, the specifications for a good society are clear'.[96] What remains is a 'serious technical problem':[97] how to produce change, who should be changed, and who should do the changing? These are the 'questions of social technology'.[98] Cartwright insists on the use of the term 'change' instead of 'nicer words' like training or education.[99] People may think the word 'change' is cold and betrays a lack of respect for human values, but precisely for that reason it will 'foster straight thinking'.[100] To keep our thinking general and basic and avoid getting entangled in all the implicit assumptions in concepts like education and therapy, we need the term 'change'.

Attempts to modify the behaviour of individuals or groups almost inevitably meet with resistance, but fifteen years of research on group dynamics had revealed how to overcome resistance to change. Cartwright summarised the results in eight principles. This 'social technology of change' works by mobilising the group as a medium of change by making use of group pressure on the behaviour of its individual members.[101] This works best (principle 1) if the people who are to be changed and those who are the agents of change 'have a strong sense of belonging to the same group', in other words, if change is not seen to be imposed from outside.[102] The problem with using the group as a medium of change, however, is that individuals are extremely resistant to changing their behaviour away from group norms (principle 5). After all, their behaviour is what it is largely because of group pressure. Therefore, the group must itself become the target of change. This can be done by making the group see the need for change (principle 6) because then, again, the 'source of pressure for change [lies] in the group'.[103] If the members of the group accept the facts that point to a need for change, ideally if they find these facts themselves in collaboration with experts, the group will bring itself to change.

Like Popper, Cartwright came very close to identifying social science and social technology. 'The fact that the subject matter of social science must be society and the fact that its methods are becoming increasingly those of experimentation have combined to produce a realization that even pure social science must inevitably be related to the problems of social technology. The experimental method implies the ability to produce change.'[104] Another similarity with Popper's views is that Cartwright saw social technology as entirely analogous to physical technology. It has produced its own 'social inventions', 'new mechanisms for the accomplishment of certain goals'.[105] Examples are role-playing,[106] buzz groups (dividing a large group into sub-groups to discuss a certain issue), process observers and post-meeting reaction sheets. To further the development of new techniques, we need, in analogy with physical engineering, 'developmental research': 'to take known principles of social science, to convert them into "new products" (such as new styles of leadership, new forms of organization, new training procedures or new systems of compensation) and then to evaluate them under field conditions'.[107] With such systematic, careful technology development, institutions may be persuaded to introduce large-scale changes while avoiding costly mistakes. It is very important, however, to realise that group dynamics is more than a 'collection of gadgets'.[108] It is first and foremost a science that specifies the principles behind these devices.

If Cartwright seemed a bit defensive, with his emphasis on science, technology and straight thinking, it was because things were not going his way. Group dynamics, he noted ruefully in 1958, had become a movement, a fad, spreading like wildfire through business, education, government, the military and many other parts of society, meeting adoration and contempt in equal measure. What worried Cartwright most about this movement is that it had distanced itself from science. This would inevitably be its downfall. The group dynamics movement would suffer the fate of all fads if it didn't keep in 'intimate contact with on-going research'.[109] Society, he intoned, 'needs an understanding of group life *on a scientific level* and this understanding can be achieved only by *research*'.[110]

The movement that Cartwright was referencing was a wayward child of his own Research Center for Group Dynamics. Its canonical origin story starts in 1946.[111] The Connecticut State Interracial Commission had asked for Lewin's help in improving the effectiveness of its staff in overcoming bias in communities. Lewin translated the problem into one of leadership and saw a chance to further develop the training techniques his group had worked on during the war. In August, adult educators, social scientists of the MIT Research Center and Harvard and public officials gathered for a two-week workshop at New Britain Teachers College in Connecticut. In true Lewinian fashion, the workshop had both a theoretical and a practical aim: the public officials would receive leadership training from the educators while the scientists studied the group processes with an eye on developing new techniques.

One evening, while the staff were discussing the observations that the researchers had made that day of the training sessions, a few of the trainees who were staying on campus and had nothing else to do asked to sit in on the meeting. Despite initial embarrassment, Lewin invited them in, and the discussion continued as if the trainees weren't there until one of them objected to the observations of her own behaviour. She proceeded to describe her own perception of what had happened in the group that day, and Lewin realised that they had stumbled upon a new source of data. Despite some misgivings of the other staff members, he invited them to come again. The following evening, a large group of trainees turned up, and from then on the evenings turned into training sessions in their own right, centred on the 'feedback', as the participants called it, that the trainees got from the observing trainers and researchers.[112] The group process itself was incorporated into the training. Group members were trained in the efficient use of group processes for creating change by reflecting on the processes in their own group as they received training.

Although it was unclear how successful the workshop had been in making the trainees more efficient leaders, the feedback sessions in particular had been so intriguing that it was decided to organise another workshop the following summer focused on group processes as such rather than on interracial relations. In one of his last contributions to group dynamics before he died, Lewin secured funding from the Office of Naval Research, and the first National Training Laboratory in Group Development took place in Bethel, Maine, in 1947. Many more would follow.

From the start, research and training had an uneasy relationship in the NTL. Even in Connecticut, the research staff had been sceptical about the evidence for actual change in the group processes.[113] The following year in Bethel the same tensions resurfaced. A list of seventy-three 'specific gripes' assembled after the workshop notes that research staff had shown a lack of involvement and enthusiasm about personal contact with the trainees. For their part, the trainees resented the many tests they had to take but even more the fact that they weren't informed about their purpose because in the interest of test validity they were supposed to remain 'naive'.[114] Yet despite these problems and general organisational chaos, the workshop made things happen. The report of one of the observers, charged with providing a group with feedback, notes the aggression towards the research programme that was expressed during the early sessions, and the irritation about the group dynamics jargon ('change agent', 'out-group', 'involvement'), which was dubbed 'Bethelesque'.[115] A rift then developed in the group between doers and thinkers, the latter especially hostile towards role-playing. But it was the role-playing in particular that gradually helped the group to turn its attention towards itself and overcome its divisions, and sensitised the members to the group processes and their own role in them. Scepticism and strife made way for enthusiastic, communal self-exploration. The final meeting 'was almost a spiritual experience'.[116] Although 'productivity' was supposed to be the central theme of the first NTL workshop, such spiritual experiences were to become Bethel's trademark.[117]

The NTL soon started to drift from 'strictly scientific procedures' and became a social movement with 'religious undertones', unconcerned with a lack of evidence for positive effects.[118] Science was less and less a consideration in the NTL; although trainers liked to point to their scientific credentials, measurement of the long-term effects of a training was seen as a pointless, reductionist exercise, and the mainstream, mechanicist theories of human psychology were rejected. The experiences spoke for themselves and could never be captured by the cold instruments of social science. The emphasis shifted away from techniques for effecting behavioural change to transformative experiences and personal change. The original technique of giving a group insight into group dynamics by making it reflect on its own processes evolved into what became known as the 'T group', the training group: 'a group devoted mainly to its own problems, to the discussion of the progress of the groups itself and of the effects of the group members on each other'.[119] The approach also went under the name 'sensitivity training' or 'encounter group'.

With the T group, the NTL acquired an increasingly psychotherapeutic character. It had already become clear during the first NTL in 1947 that the feedback process could be emotionally disturbing, and it was regretted that there were no counsellors present to help trainees who were upset by the experience. In 1949, clinically oriented trainers were added to the staff, but within months they had taken control of the T groups, and the original, Lewinian trainers had to step back. In 1950 the NTL became an independent, year-round institution funded by the Carnegie

Foundation. Dorwin Cartwright, as the director of the Research Center for Group Dynamics, still had a seat in the review and coordinating committee,[120] but at the Research Center the feeling was growing that its association with the NTL was a threat to its academic credibility.[121] When Cartwright in 1958 predicted the downfall of the group dynamics movement if it didn't return to its scientific roots, he was wrong, however. The transformative experiences of the T group would only get more popular over the next decade. Even Alfred Marrow was a devotee, and he considered the fact that many hard-headed businessmen used the method sufficient proof of its success.[122]

Science and Sensitivity

The group work that Lewin and his students did in the early 1940s spawned widely divergent ideas and practices. Dorwin Cartwright envisioned a future in which the inventions of group dynamics would be put to use in constructing a better society. This was engineering work but in the social realm rather than the material world. It was the development of new mechanisms of social management in a cautious process of trial, evaluation and redesign with the generally agreed-on specifications of a good society as a guiding ideal. The similarity between Cartwright's social technology and that of Karl Popper is obvious, although neither ever mentioned the other. Popper would have objected to the object of Cartwright's technology, to influence individual behaviour and mould social relations, but he would have approved of the general idea: that society has to be governed in accordance with scientific principles and that democratic governance is a social technology.

The T groups of the 1950s and 1960s, on the other hand, although equally tracing their origins to Lewin, represented quite a different attitude, gradually turning more and more against a technological view of human beings. The kind of science that Popper and Cartwright represented was rejected in favour of an intuitive approach that emphasised emotions and the here and now rather than ratio and instrumental considerations. The group was no longer focused on an outside task such as changing food habits or achieving higher production but on the immediate experience in and of the group. The T group was a way to discover yourself without any immediate utilitarian aims. The many business executives who came to the NTL in the 1960s to become 'Bethelised' did not learn new management techniques; they were sensitised, able to see themselves as others saw them.[123] That's what made them better managers.

Between Cartwright's technology and the T group's sensitivity, the tact that Bavelas had embodied fell out of view. Cartwright never mentioned anything resembling tact when he wrote about social technology. In his conception, social technology was a disembodied collection of principles and devices that could be developed and tested just like physical machines and artefacts. Cartwright did note that the techniques of group dynamics weren't 'gadgets' or 'tricks', but that was to emphasise the link with the science behind them, not their embodiment in a tactful change agent. In T groups, tact was not a virtue either. Whereas in the earlier work, that of Bavelas in particular, the way to bring people to change was typically subtle and indirect, and often the aim of the training was equally to instil this tact in the trainees, the T groups cultivated the confrontation. Rather than gently guiding groups or individuals to new behaviour, in T groups participants would be encouraged to say exactly what they thought of each other. Spontaneity and authenticity took the place of tact, both as a means and as an aim. Leading a T group was not subtly manipulative but openly non-directive. The group leader's refusal to lead forced the group to confront itself. Managers who had been sensitised by the T group experience were supposed to be open and nurturing leaders, not tactful change agents.

8 Mind Control

> It is easier to manipulate the mind of others than to avoid doing so.
>
> —Joost Meerloo[1]

In the 1950s, Western societies were gripped by the fear that the Communists had developed techniques for turning people into machines. The first indications that the Soviets possessed mind control methods had emerged before the war.[2] During Stalin's purge of the Communist Party between 1936 and 1938, critical party members had gone on trial accused of treason and other crimes. Motion picture images distributed over the world showed them admitting their crimes with apparent enthusiasm, denouncing themselves as traitors, praising the party for its wisdom and righteousness and begging for the death sentence.

Soon after the war, the trial of the Hungarian Cardinal József Mindszenty was a new, particularly powerful trigger of the fear of Communist thought control. Mindszenty, a determined and vocal critic of the new Communist rulers in Hungary, had been arrested in December 1948 on charges including treason and conspiracy. In a show trial a few months later, a visibly dazed Mindszenty confessed to all the charges, in monotonous, mechanical phrases. It was an eerie, deeply disturbing sight. A year later, in the same courtroom, before the same judge, American businessman Robert Vogeler confessed to espionage and sabotage and apologised remorsefully for his subversive acts. His wife and friends were baffled by his uncharacteristically calm, impassive demeanour. He didn't seem himself.

Mind control is the dream of efficient and effective control over people's behaviour turned into a nightmare.[3] In the 1930s and 1940s, the nightmare was described in several novels. Arthur Koestler's *Darkness at Noon* was a dramatised account of the show trials during Stalin's purge. *Brave New World* by Aldous Huxley and *Nineteen Eighty-Four* by George Orwell depicted worlds in which science, including behavioural science, had made it possible for the state to control people's thoughts and emotions, and control them to such an extent that – this was perhaps the most frightening part – people accepted such control as normal and desirable.

The trials of Mindszenty and Vogeler seemed to confirm that these novels depicted something real. Apparently it was really possible to manipulate other people as if they were machines, and the Communists were using these techniques for evil purposes.

The journalist Edward Hunter gave the nightmare a name that stuck: brainwashing. It was supposed to be a set of techniques that could make a person sincerely believe and profess things that weren't true, but from the start, fact and fiction were also intricately entwined in the discourses around brainwashing.[4] The term itself appears to have been created with propaganda aims in mind. 'Brainwashing' first appeared in internal documents of the Central Intelligence Agency (CIA) in January 1950, nine months before Hunter first used the word,[5] and Hunter, it later turned out, was himself a CIA operative working undercover as a journalist. It is not unlikely that the CIA popularised the concept of 'brainwashing' via Hunter in order to get government approval for its own mind control projects[6] or as Cold War propaganda aimed at American citizens.[7] It is certain that the CIA later exploited the brainwashing scare for both these purposes.

For these reasons, the historian Charles Young has flatly labelled brainwashing a 'hoax' perpetrated by the CIA, but this is too simple.[8] The top brass at the CIA did not treat brainwashing consistently as a propaganda ploy. At the same time that it set in motion the exploitation of the scare that it had itself created, the CIA also put scientists to work to investigate the Communist techniques and their effectiveness, referring to them as 'brainwashing'. It also funded experimental research to develop its own brainwashing technology. The CIA seems to have operated on the principle that even if Communist mind control wasn't real, perhaps it could be made real by American scientists. Historian Timothy Melley has noted these complications in an insightful exploration of the entwinement of truth and fiction in 'brainwashing', but he spoils it in my view when he calls it 'a strategic fiction' produced by (parts of) the CIA.[9] Firstly, the fact that the CIA coins a term does not necessarily make it fiction. Deception may be an important part of the CIA's operations, but as an intelligence agency it is also intensely concerned with the truth. Secondly, although Hunter certainly played up the frightening power of brainwashing techniques, I have no doubt he believed most of what he was describing. 'Brainwashing' was not *merely* strategic and fictional, although it was those too.

The problem with the epithets 'strategic fiction' and 'hoax', I will argue, is that they ascribe too much control to the CIA. 'Brainwashing' had a life of its own in which the agency certainly had an influential role but that it never had full control over. The CIA was under the influence of 'brainwashing' as much as it tried to control and exploit it. On a different but connected level, the same is true of the practices that the term 'brainwashing' referred to. Whereas brainwashing was first depicted as the act of a powerful, controlling brainwasher against an incapacitated victim – 'the calculated destruction of men's minds', as Hunter put it – during the 1950s a more complicated picture gradually emerged.[10] The academic researchers who studied the Communist interrogation and thought reform practices for the CIA, the Air Force and Army concluded that what had been called 'brainwashing' involved interpersonal processes that were only calculable to a limited extent and in which the interplay between the agency of the brainwasher and of the victim was the central dynamic. 'Brainwashing', I will argue, came to be seen as tactful and rhetorical.

Brain-Washing and Menticide

Hunter first used the term 'brain-washing' in 1950 in an article in the *Miami Daily News*. The article did not reach a large audience, but his book *Brain-Washing in Red China*, published a year later, became a bestseller.[11] Its cover promised an explanation of 'the mystery behind the voluntary "confessions" of the innocent' and an exposé of 'the terrifying methods that have put an entire nation under hypnotic control'. That nation, China, had come under Communist rule in 1949. Hunter, who at the time was based in Hong Kong, had spoken with a number of refugees who told him of the thought reform programme that the Communists had introduced as soon as they had taken power. It was referred to by the Chinese, or so Hunter claimed, as 'brain-washing' and 'brain-changing'. The former was a fairly simple procedure of indoctrination; the latter was 'immeasurably more sinister and complicated'.[12] Brain-changing involved not only hypnotism but also drugs and 'cunning pressures that plague the body' in order to remove specific memories from a person's brain and fill the gaps with ideas the authorities wanted him to 'remember'.[13] Mindszenty, for example, was a victim of brain-changing. It was as yet beyond the capabilities of the Chinese, but in time they would use it too.

Hunter did not explain why the Chinese public had coined a phrase, brain-changing, for a practice that did not yet exist in China. Whether the term 'brain-washing', 'xi'nao' in Chinese,[14] was actually used in China is also dubious. Robert Lifton, who like Hunter interviewed Chinese and Western victims of Chinese re-education practices, never heard it from his informants, although they did use metaphors like 'cleansing yourself'.[15] Hu Ping, who experienced the thought reform campaigns first-hand, never came across 'xi'nao', although the politically charged 'xi'zao' (bathing) was in use.[16] Even in the term 'brainwashing' itself truth seems to have been mixed with a liberal dose of fiction.

Although simpler and less sensational than brain-changing, brainwashing had its own horrors. One of Hunter's contacts, a student, described how he had been sent to a revolutionary university directly operated by the party after he had publicly questioned Russia's role in China. The regime in the revolutionary university was harsh, with insufficient coal to heat the rooms in winter and a very frugal diet. Students were expected to labour long hours in the field tending crops. The actual re-education happened in lengthy lectures on the theory of historical materialism but especially in 'democratic group discussions' and 'self-criticism sessions'. The outcome of the discussions and self-criticism had to correspond to official doctrine, guarded by the party members present, but the students were supposed to discover it of their own accord and accept it voluntarily. What got them to that point was the exhausting regime, the threat of being sent to an even harsher school and the knowledge that their families were under surveillance. With their defences worn down, students could no longer resist the ideas that were constantly impressed upon them and that they were made to repeat themselves. It was impossible to escape 'the psychological pressure exerted by this skilfully devised technique of creating a hypnotic state of fatigue and forcing a person, while in this trance, to repeat again and again, lengthily and in his own words the political dogma demanded of him'.[17]

For Hunter, democratic group discussion was the main technique of Chinese brainwashing, 'the machinery for the intellectual conquest of the Chinese'.[18] In Hunter's description, democratic group discussion seems a sinister parody of the work of Lewin and Bavelas, with its goal of voluntary agreement with what had in fact been decided beforehand and the skilful manipulation of the discussion process by the teacher. Hunter appears not to have been aware of their work,[19] but in the opening chapter he does describe his alarm when he realised the parallels between the stories of the Chinese student and several experiences he had had back in the USA. Suddenly, it all fell into place. During one incident, a psychiatrist had explained to him that a patient could only be cured if he brought forth and accepted, voluntarily, the truth that was hidden in his own mind. Hunter had also witnessed a stage hypnotist implant a post-hypnotic suggestion in the mind of a member of the audience during a show. The man had suddenly risen from his seat half an hour later to walk along the aisles 'selling fish'. Hunter had been struck by the thought that, if this could be done for fun in the theatre, it could be done anywhere else to achieve any other goal. Suddenly he realised that '[t]he whole field of psychology has broadened to embrace everything that influences thought and attitude'.[20] Psychology had learnt how to manipulate and control the brain.

In the same year that Hunter published *Brain-Washing in Red China*, the Dutch-American psychiatrist Joost Meerloo introduced a term that was at least as frightening: 'menticide'.[21] Meerloo had escaped occupied Holland during the war and had been employed by the Dutch government in exile as a psychological advisor. In the winter of 1944–45, known as the 'hunger winter' in the Netherlands, Meerloo was charged with organising food aid to the Dutch population.[22] His interest in mind control dated from the time of Stalin's purges in the 1930s and had been rekindled when he read Koestler's *Darkness at Noon* when in exile. When the French magazine *Esprit* in 1951 published the results of a poll of psychiatrists and psychologists about the possibility of 'direct psychological intervention in an integrated individual', Meerloo decided to put his own views on paper, drawing among others on his own experiences during the war.[23]

With a combination of academic precision and dramatic flair Meerloo defined menticide as 'an organized system of judicial perversion and psychological intervention, in which a powerful tyrant synthetically injects his own words and thoughts into the minds and mouths of the victims he plans to destroy'.[24] In Meerloo's eyes, menticide is deadlier than death itself. It is a threat 'far worse than genocide', 'more deadly than lethal weapons'.[25] Menticide destroys a man's mind and free will and turns his thought processes into servile mechanisms. Meerloo called on his fellow psychiatrists to act because psychiatry and psychology were implicated in the crime of menticide. As our understanding of the human mind has grown, so has the realisation that it can be 'kneaded like putty', and tyrants have not hesitated to draw on this knowledge.[26]

Meerloo distinguished social and individual menticide. The seduction of the German people by Hitler and Goebbels is an example of the former. It works by spreading fear and suspicion via artificial rumours and mass propaganda and by systematised suggestion. Meerloo discussed individual menticide more extensively. Here, physical brutality is an important element. The victim is inadequately fed and deprived of sufficient sleep. It may involve torture – Mindszenty had been forced to stand for days on end, resulting in excruciating pain – but the strong minds of people like Mindszenty cannot be broken by pain alone. It is 'the horror from within' that really breaks the resistance of the victim.[27] Mental torture is the essence of individual menticide. Hungry and exhausted, the prisoner is tormented by the unsteadiness of his mind, by the need for companionship and simultaneously the need to be left alone. 'Hidden guilt feelings' undermine his rational knowledge of his innocence, and his dependence on the interrogator reduces him to the state of a helpless infant.[28] If he proves too resistant, he is given narcotics to further unhinge his mind. Once his resistance is broken, the interrogators begin to bombard him with constantly repeated questions and accusations. Eventually, the prisoner gives in to the 'need for infantile self-preservation', acknowledges his dependence on his persecutors and confesses to crimes he hasn't committed, thus covering his unconscious feelings of guilt.[29] Now starts the final phase, in which the victim is made to train himself by having him reread and repeat his confessions until he is convinced of his fabricated crimes. At last he is ready for his trial.

Like Hunter, Meerloo stressed the hypnotic effect of repetition, but he added that it also conditions the victim. The second phase of menticide, when the victim learns to accept and speak new 'truths', is a form of training, '[m]uch like an animal is conditioned to perform tricks'.[30] All in all then, menticide is a mixture of psychoanalysis, conditioning, hypnosis and pharmacological interventions: techniques of psychology and psychiatry used for evil purposes. To 'immunize' people against menticide we may begin by deploying the forces of psychology for good.[31] Just as psychotherapy strengthens the inner resistance of neurotic patients, new ways of rearing children are beginning to produce people with stronger egos. In the future, people will be better able to withstand outside pressure. But that is not enough to resist the destructive power of menticide. We could train people in autohypnosis, 'for the hypnotic state can offer immunity to pain, hunger, and psychological intrusion', but we cannot expect everyone to become a yogi.[32] Psychoanalysis, to become aware of 'early guilt and defense mechanisms', is another way to build resistance, but analysing everyone is as yet impractical.[33] For now, we must rely on social protection. Meerloo had his hopes pinned on the United Nations, which should draw up rules against psychological attack. 'Mankind should be guaranteed the right *not* to hear and not to conform and the right to defense against psychological attack in the form of perverted mass propaganda and individual mental torture.'[34] And it all begins with awareness of the danger and giving it a name. Coining the word 'menticide' is the start of the fight against it.

Although they do not appear to have been aware of each other at the time, Hunter and Meerloo described brainwashing and menticide in largely similar terms. Both saw it as a psychological procedure that was facilitated by physical hardship and the exhaustion and confusion it creates. Repetition was a crucial element, especially repetition by the victim himself of the ideas and phrases that were expected of him. Repetition creates a state of hypnosis and autohypnosis, which makes the victim accept as his own what has in fact been forced upon him. Meerloo added that the victim is conditioned in this way. The result is an individual or an entire nation reduced to the state of a mere mechanism, bereft of free will. Noteworthy is also the fact that both saw a similarity between brainwashing or menticide and psychotherapy. Finally, they put the labels 'brainwashing' and 'menticide' on a wide variety of practices – mass propaganda and Chinese groupwise thought reform as well as Soviet individual interrogation – and they warned that, although the main danger lies elsewhere, in the Communist world, we in the free world are also under threat. Here in the West the knowledge and techniques of psychology and psychiatry are also starting to be used to control the human mind.

Propaganda

The Korean War (1950–53) made Communist mind control an acute threat. In March 1952, China accused the USA of biological warfare in Korea. Week after week the accusations were repeated and immediately denied by the Americans. Then the Chinese produced their trump card: public confessions by two captured US pilots. Another thirty-six followed. China had scored a major propaganda coup. The American government knew the accusations were false, nor did the American public lend them any credence, but the confessions did seem sincere.[35] Relatives of American prisoners of war, moreover, had been receiving letters from Korea in which their loved ones expressed sympathy for the North Korean cause and criticism of the unjust war that the UN and USA were waging against the brave people of Korea. Were these 'confessions' and Communist sentiments the result of brainwashing, or were they signs of collaboration?

The reaction of the American public and press oscillated between 'pity or punishment', as one press columnist put it: between forgiving the POWs as victims of Chinese brutality and condemning them as traitors or unmanly weaklings.[36] The American government was, and would remain, equally ambivalent. Completely exonerating all collaboration was out of the question, but putting POWs on trial who had suffered at the hands of the Communists did not look good either. It was on the one hand a propaganda matter: to counter the germ warfare accusations and restore America's image abroad and to defend the reputation of the American armed forces as a formidable opponent. But it was also essential to examine what had happened, what had caused the fliers and soldiers to betray the cause they were fighting for. Image and truth were simultaneously at issue, and with regard to both these aspects 'brainwashing' played an important role in the American response.

On the eve of 'Little Switch', the first, limited exchange of prisoners between China and the UN/US forces, the American government decided it was time to confront the collaboration problem head on. Allan Dulles, the director of the CIA, convened two meetings of the Psychological Strategy Board (PSB) to discuss how to respond to the confessions of the pilots and how to deal with the returning POWs and in particular the collaborators among them. The PSB had been established under Truman to coordinate psychological warfare and included representatives of the State and Defense Departments and the CIA, on this occasion supplemented with representatives of the 'psywar offices' of the Army and Air Force. The main goal of the meetings was to explore the possibility of forming a group of doctors and/or scientists to 'examine the facts and issue a statement regarding the feasibility of brainwashing procedures'.[37] Either Dulles did not know the term 'brainwashing' was a CIA creation, or he was simply being careful not to let other officials in on the secret.[38]

The study group was to be presented with the available evidence as well as 'names of persons who have been subjected to brainwashing treatment in China', but Dulles also noted that '[a] quick search in CIA shows little scientific evidence to support brainwashing.'[39] There *is* overwhelming evidence, he added, that with enough time it is easy to force any person to make a false statement, provided that the time at which the statement is to be made can be controlled, and 'the individual is not required to continue his willingness to lie for a long period'.[40] Twisting someone's arm, literally or metaphorically, only works for so long. Finally, Dulles mooted the possibility of having the scientific committee 'consider the length of time brainwashing may be effective'[41] (this being what apparently set it apart from known methods of forcing someone to make false statements) but advised not to spend too much time on it. Dulles, it seems, was sceptical about the claims made about brainwashing, perhaps because he knew those claims originated in the CIA, but he was on the other hand not dismissive about them. They were worth looking into.

The PSB itself set up a working group to assist the future scientific committee. Within two days, this working group sent Dulles a report. Reflecting the PSB's propaganda perspective, the report focused on how false confessions and POW collaboration should be presented both at home to the American public and internationally. It repeated the recommendation to create a scientific committee to investigate brainwashing but also warned that '[w]e should be careful not to credit the enemy with a "miracle-weapon"'. The term 'menticide' should equally be avoided, since it is 'too strong a word or too frightening for the American public'.[42] The group was looking for a better catchword. In a press release included in the report, the working group made a start allaying the public's fears of the Communists' miracle-weapon. The false confessions were simply the result of feeding the prisoners of war endlessly repeated misinformation. However, the word 'coerced' is deleted in the document: if the confessions were the result of coercion, there would be no room for holding POWs culpable.

A month later however, when the American POWs who were exchanged during Little Switch were about to come home, the PSB decided to play the brainwashing card. In a memorandum the PSB urged that 'the American public be prepared, beginning immediately, to face the grim realities of the brain-washing process'.[43] The cooperation of the press must be enlisted to explain the problem, and, most importantly, the families of POWs must receive a letter, preferably from the chief of staff himself, informing them 'factually and sympathetically' of what had happened but nonetheless making it clear that 'this mental aggression is one of the new horrors of modern warfare as the Communists wage it'. In addition, the families should be visited by a high-ranking officer who was especially instructed 'on how to explain the brain-washing problem'.[44]

CIA director Dulles contributed to the campaign with a rare public address to Princeton graduates. He used the starkest terms to warn them of the 'brain warfare' that the Soviets were waging. Dulles knew the psychological mechanism by which it worked. Inspired perhaps by Meerloo, he claimed the Soviets aimed to 'condition the mind so that it no longer reacts on a free will or rational basis but responds to impulses implanted from outside'.[45] Their 'brain perversion techniques' are 'subtle and abhorrent' and serve in a vast experiment to change the mind of an entire people.[46] The brain of every Russian citizen is turned into a phonograph playing a disc that the Communists had put on its spindle. Dulles even suspected the Russians of possessing a miracle weapon: a 'lie serum' with which they 'create new brain processes and new thoughts which the victim, parrot like, repeats'.[47]

Towards the end of his speech Dulles gave a much more prosaic description of brainwashing. Little is known about the details of the brainwashing process, he said, but some victims had escaped and lived to tell about it. The Russians, it appears, broke their prisoners by depriving them of sleep and submitting them to months of nightly interrogations, usually with a bright light shining in their face. Sometimes the prisoners were made to stand in awkward positions for a long time, long enough to make it extremely painful. There was no mention of a lie serum. It was a strangely sober ending of a speech that was otherwise so florid.

This first propaganda campaign around brainwashing quickly became a shambles.[48] Repatriates who had been diagnosed as brainwashed publicly denied they had ever harboured Communist ideas. The press, public and congressmen turned against the Pentagon, accusing it of a smear campaign against the soldiers. Some officials stuck to the plan and pushed the link between collaboration and brainwashing, but other officials downplayed it. An internal PSB review concluded that the campaign had been an incompetent failure. As Young has noted, the events surrounding the repatriates of Little Switch show that the association between collaboration and brainwashing was there, but it was not solid. 'There was fear of red zombies, but it was balanced by scepticism and sympathy for the soldiers.'[49] Nevertheless, later in the year, when the rest of the POWs came home at 'Big Switch', a new propaganda offensive centred on brainwashing was mounted. This campaign was more successful: it succeeded in amplifying what Hunter and Meerloo had started into a full-blown brainwashing scare and strengthened the association between brainwashing and Pavlovian conditioning.

The campaign was the brainchild of the Operations Coordinating Board (OCB), the PSB's successor, and was outlined in the 'National Operations Plan to Exploit Bacteriological Warfare Hoax, Mistreatment of Prisoners of War, and Other Atrocities Perpetrated By Communist Forces During the Korean War' of August 1953. An important goal of the campaign remained 'to destroy and counterexploit the Soviet bacteriological warfare myth'.[50] To that end the OCB considered it 'necessary to note that Soviet methods of psychological coercion are capable of compromising most individuals whom they are determined to break'.[51] (If any victims of such coercion were still to be tried for collaboration, the OCB suggested it would be best to keep publicity to a minimum.) Secondly, the analysis of the Soviet methods should be completed at short notice and the results exploited in counterpropaganda and diplomacy. Meanwhile, American citizens and other people of the free world should be educated about the mistreatment of POWs, including '"brainwashing", and comparable tactics'.[52]

The OCB mounted a media offensive that included television and radio programmes and magazine and newspaper articles featuring selected experts and POWs, all tightly coordinated by the OCB to avoid another debacle like after Little Switch. 'The nation itself became the target of psychological warfare.'[53] The rest of the free world was first educated about the Communist mistreatment of POWs in an address to the United Nations by Dr. Charles W. Mayo, son of Mayo Clinic founder Charles H. Mayo. As 'alternate representative', appointed by president Eisenhower, Mayo told the Political and

Security Committee of the United Nations that the captive pilots had made their germ warfare confessions under extreme duress. Their Chinese and North Korean interrogators were trained and supervised by Soviet personnel. They used tortures that were 'calculated to disintegrate the mind of an intelligent victim', such as sleep deprivation, a minimum of food, and solitary confinement.[54] When the captive's resistance was broken, he was trained to produce the required confession, using 'the conditioned reflex', 'the same technique which the famous Soviet biologist Pavlov used in his experiments on dogs and rats'.[55] The Soviet regime, he added, 'has used this same technique against its own people to dislodge them from their traditional reverence for the Almighty and from their aspirations toward freedom; and to force them willy nilly into the Communist slave pattern'.[56] That so many soldiers 'somehow continued to act throughout like men',[57] and that some of the pilots refused to yield altogether, was proof that 'the spirit of man can run deeper than the reflexes of Pavlov'.[58]

The OCB made sure that Mayo's statement was widely distributed both abroad and in the United States. Although Mayo had not used the term 'brainwashing' in his statement, the idea that brainwashing worked by conditioning rooted itself in the public imagination. As historian Susan Carruthers notes, '[i]t would be hard to overstate the ubiquity of Pavlov in contemporary analyses of brainwashing'.[59] 'Pavlov' explained how brainwashing worked and confirmed the suspicion that the Soviets were the evil masterminds behind the events in Korea. It became common knowledge that conditioning, often in combination with hypnosis (conditioning under hypnosis, for greater effect), was the mechanism by which victims were turned into automatons, puppets controlled by an evil, usually Communist master.

The brainwashing scare that the OCB's campaign whipped up firmly entrenched the concept in popular culture. Brainwashing became a staple of movies and books in which the era's preoccupation with the struggle between individual freedom and social control was worked through.[60] Novels and movies like *The Manchurian Candidate* used narratives around brainwashing to speak to the general anxiety about the loss of individuality in a mass society and the influence of the mass media. Hypnosis and conditioning were standard elements in these stories, signifying the loss of autonomy, the submission to manipulation and the resulting degraded, mechanical state of being that were associated with brainwashing and modern society in general. Hunter and Meerloo, the two authors who had first sounded the alarm about brainwashing, also contributed to the debate and put the phenomenon in the wider context that readers now expected. Both gave Pavlov a prominent role in their stories.

Apex

Hunter had been alerted to the relation between brainwashing and the work of Pavlov by a brain surgeon and psychiatrist of his acquaintance, Dr. Leon Freedom. Freedom was a defense witness in the case against Claude Batchelor, one of the POWs who 'chose China' and defected after the Korean War. Batchelor had changed his mind at the last moment and was court-martialled. Dr. Freedom had not examined Batchelor in person but gave his expert opinion that Batchelor suffered an 'induced political psychosis' as a result of brainwashing. The court was not impressed and Batchelor was convicted, eventually serving four and a half years.[61]

Alerted by Freedom, Hunter had discovered that there really was a connection between Pavlov's experiments and the Korean POW camps.[62] What the totalitarian state aims for, he realised, 'is no less than the insectivization of human beings', and conditioning gives them the tool to do this.[63] A film clip of a young man being conditioned with cake and biscuits as reinforcement was shocking proof of the Communists' perverted ideas about human nature. These techniques gave the Communists remote control over their subjects: 'Any human activity, from the flow of saliva to an embrace or a murder, could be clinically predetermined in politico-medical laboratories by connecting it with a shouted or written slogan, a hand signal, a smear word, or the color of a man's skin.'[64]

Hunter's account of how brainwashing worked was a jumble of theories and concepts. Dr. Freedom supplied him with a theory that described brainwashing as psychiatry in reverse: a set of techniques that exploit psychodynamic, neurological and conditioning processes to create a pathological condition that brings a man to a false confession. Brainwashing was 'corticovisceral psychiatry'.[65] However, Hunter also saw an analogy with evangelism, in particular the centrality of the 'confession' as a submission to the authority of the party, and he noted that much of the process is persuasive and argumentative. The Chinese were devilishly clever in debate. All in all, it is 'a very intricate manipulation, more like a treatment than a formula'.[66] But, he insisted, it is more than mere persuasion, indoctrination or a return of the Inquisition because it is scientific. 'The Inquisition had no Pavlov and was not thought up in a physician's laboratory.'[67]

Fortunately, man is more than an animal that can be conditioned like a dog; he has reason. As long as he keeps his reason, judgement and free will intact, his future is safe. If, however, he lets 'the divine traits in him to be overcome and his reasoning power – his judgment and free will – to be atrophied, he could be made into a demon, a puppet'.[68] Most of the book consisted of interviews with the 'men who defied it': the Korean POWs who had been subjected to brainwashing but hadn't cracked. Their stories taught Hunter that brainwashing is not irresistible, as it was often made out to be. The Reds themselves cultivated this image because the defeatism it bred in their victims softened them up for brainwashing. Confidence was a prominent trait of those who had been able to resist the Communist brainwashing. Even more important were strong convictions, be they religious, political or moral. To resist the Reds' double talk and 'upside-down logic'[69] it was vital to have firm beliefs to hold onto, primary among them the conviction that there is no truth in Communism. Liberalism and open-mindedness were fine, but anyone in his right mind should see that Communism is wrong and not let go of that idea. 'The mature thinker's approach to communism is that it is evil, not partly evil but all evil.'[70] It was imperative that this truth be impressed on everyone from a young age. Everyone should be given 'mental-survival training' and learn about the dangers of brainwashing and Communist thought so they wouldn't get lost in the 'ideological jungle' of Communism and would be able to defy the brainwashers should they ever fall into their hands.[71] Victory in the battle against the Reds, this 'ultimate conflict for the minds of the people of the earth', depended on it.[72]

Joost Meerloo also rejoined the fray.[73] When the Marine Corps in 1954 held a court of inquiry regarding the conduct of Col. Frank Schwable, one of the pilots who had 'confessed', Meerloo was a defence witness. Schwable, he told the hearing, could not be held responsible for his false confession. Menticide could reduce anyone to 'a mechanical imitator of his tormentors', just like Schwable.[74] Meerloo now also emphasised the Pavlovian angle. In a new article on brainwashing, he claimed the Communists were abusing Pavlov's findings to create 'a new society of conditioned human insects'.[75] The Soviets had created a total system of conditioning that first gets 'the required response from the nerve cells, then the control of the individual, and finally the control of the masses'.[76] The soldiers in Korea had been subjected to the same kind of conditioning, using a wide array of negative stimuli (hunger, fatigue, boredom, confusion, etc.) and a few positive ones (food, mainly) to create the kind of responses that were required. Such conditioning, Meerloo added, is only possible when the individual is overcome by fear and hopelessness and when his captors succeed in inducing feelings of guilt in him – Freud was never far in Meerloo's thinking. And, of course, the constant repetition of stimuli induced hypnosis in the prisoners. What had been perpetrated on the POWs in Korea was also done on the masses in totalitarian states and, Meerloo warned, in our own society if we are not careful.

The 'automatizing attitude', the 'intrusion of a mechanizing action and technique'[77] threatens here as well. Our schools are in danger of producing 'conditioned adepts' rather than 'free-thinking students'.[78] Popper would certainly have agreed.

Two years later, Meerloo published his magnum opus, *The Rape of the Mind: The Psychology of Thought Control, Menticide, and Brainwashing.* In *The Rape of the Mind*, published in England under the slightly less arresting title *Mental Seduction and Menticide*, Meerloo further developed the themes he had introduced in his 1954 article. Brainwashing, he warns again, is an extreme symptom of a malaise that is threatening all of modern culture all over the world: mechanisation. We are in danger of letting the machines, the tools that we created to gain more power, gain power over us. Totalitarianism is, at heart, nothing other than this technological dominance. The mechanical thinking of the Soviets, who mistake man for a robot and abuse Pavlov's discoveries to create a society of mindless followers of a tyrannical leader, is making headway in the free world as well. 'Menticidal forces' are threatening us here too.[79] The addiction to television, the influence of advertising, the growth of bureaucracy, the focus on examination in education and the replacement of the human brain by mechanical computers are all signs of encroaching totalitarianism. The machinery outside us is turning our minds into machines. Technology 'has forced people into automatic responses, into rigid patterns and destructive habits'.[80]

Meerloo was particularly scathing about television. Not only is television habit-forming and does it lead to 'precociously sexual and emotional turmoil' in youngsters,[81] it is also a prime example of the way modern technology connects the world so seamlessly to our senses that we no longer reflect on it. Propaganda and advertising can invade our minds unnoticed because television presents the world ready-made in attractive images. Television, cars and other modern devices are so convenient that we no longer need to think or struggle, leading to 'mental and physical atrophy'.[82] Man needs challenges to become strong and autonomous.[83]

For Meerloo, brainwashing was the essence of totalitarianism and, if we're not careful, of modern society generally; in fact all social processes continually risk deteriorating into brainwashing. Each individual is immersed in a web of influences that constantly impinge on his unconscious, triggering hidden fears and wishes, suggesting ideas and directing his feelings, all without his awareness. In each person, hidden from his 'free alert mind' lurks an 'automaton' that is being manipulated by the forces around him.[84] All of us, moreover, are manipulators, driven by 'deeply hidden criminal and hostile thoughts and feelings'.[85] But we are all, at the same time, willing victims, with a need for punishment and an unconscious wish to give in and comply. In brainwashing, a 'mysterious masochistic pact' is established between interrogator and victim, but really all social interaction risks degrading into such a relationship; 'general mental coercion may belong to every human interaction'.[86]

What allows us to transcend such a society of generalised coercive interaction is our individual, conscious rationality. To understand and be aware of processes of coercion is to enable ourselves to resist them. Soldiers need to be educated about brainwashing in situations of war and ordinary citizens about the coercive aspects of everyday social life. 'We cannot be enough aware of the continual coercion of our senses and minds, the continual suggestive attacks which may pass through the intellectual barriers of insight.'[87] We must raise our children to become autonomous, critical thinkers who dare to be creative and spontaneous. We must foster intelligent conversation rather than mindless consumption, and we must challenge ourselves rather than seek conformity and convenience. Individualism is an absolute requirement because 'every step in development needs isolation, needs inner conversation and deliberation and a reviewing with the self'.[88] Only if we're able to step back from the social process and reflect on it can we safeguard our autonomy and moral integrity. 'Moral culture begins and ends with the individual.'[89]

The hope for salvation lies in psychology, which for Meerloo is more or less synonymous with psychoanalysis. Although psychology has been abused for totalitarian ends, it is inherently a democratising force that lays bare an individual's inner tensions and enables him to be free and responsible, autonomous and 'self-growing'.[90] Psychology helps the individual recognise the forces that influence him and thus gives him the power to resist them. Psychological insight allows the individual to transcend the web of social influences and become a truly moral member of society, one whose actions follow the rules of society that he has freely accepted.

In Meerloo's work, the brainwashing scare found its apotheosis. Its combination of Pavlovian learning theory and psychoanalysis made both the mechanism and the motive behind brainwashing ubiquitous. Below the surface, all of social life consists of processes of mutual influence that constantly risk turning into brainwashing because, below the surface, all individuals have both the desire to control others and to yield to them. The unconscious is the domain of learning mechanisms that can make us vulnerable to unwitting influence, and it is the seat of hidden desires and feelings of guilt that give us the motive to exploit this vulnerability or give in to it. Thus, brainwashing is a danger lurking in every individual and every social interaction.

Brainwashing, moreover, is also happening on a cultural level. The growing dominance of technology in our culture is beginning to brainwash our entire society. Totalitarianism threatens wherever the automatic, mechanical aspects of man and society are developed at the cost of freedom and rationality. The greatest flaw of Communism is not its economic theory but its belief in a technological utopia: a society that doesn't merely use technology as an instrument to master the material world but allows it to control human relations. Technology is starting to overwhelm Western societies as well, where the ever-present machinery around us forces us into mechanical responses, where radio and television bring the influence of advertising directly into the consumers' unconscious, where election campaigns are masterminded by 'public opinion engineers'[91] and government is being mechanised into bureaucracy. Meerloo's standpoint is more extreme than that of Popper: whereas the latter endorsed social technology but rejected human engineering, Meerloo warns against any blurring of the boundary between social relations and technological processes. Man must always keep the technological at arm's length.

The Rape of the Mind and its author were highly praised in the *New York Times* by psychologist John Dollard. He admired the way Meerloo, 'social scientist extraordinary', combined psychoanalysis with social science (which was also Dollard's own project), but he was even more impressed by the width of Meerloo's perspective. Dr. Meerloo is a 'passionate spokesman for the democratic practice of life as a general human goal, not merely as a device for beating off the totalitarians . . . everyone should know him'.[92] Social psychologist Edgar Schein was more reserved. He thought *The Rape of the Mind* was a stimulating, thought-provoking book but deplored its lack of evidence and cohesive argumentation. Although it purported to be a scientific discussion, the book was 'essentially a polemic'.[93] Schein was one the scientists who had studied Soviet and Chinese interrogation and indoctrination methods for the American government, the Army in Schein's case.[94] Their reports were published around the same time that the books of Hunter and Meerloo appeared, but their message was quite different. Whereas Hunter and Meerloo, following the OSB propaganda campaign, had inflated brainwashing to the status of number-one threat to Western civilisation, Schein and his colleagues unanimously tried to deflate it.

'Brainwashing'

In the reports of these scientists, 'brainwashing' usually appeared in quotation marks, indicating their distance from this popular concept.

They preferred more academic-sounding terms like 'communist control techniques' or 'coercive persuasion' and poured scorn on the sensationalism of popular writers and the hysteria of the public. When considered from a dispassionate, scientific point of view, they reassured their readers, these techniques were nothing out of the ordinary. There were no 'occult methods'[95] or 'mysterious devices'[96] involved. Brainwashing was psychology, not 'demonology'.[97] Psychologist Albert Biderman, who had been hired by the US Air Force to investigate what had moved captured pilots to make false confessions, reported that the Chinese techniques were nothing special. The single most spectacular finding, he wrote, is 'that essentially there is nothing new or spectacular about the events'.[98] People had reacted incredulously to this finding, he added. Lawrence Hinkle and Harold Wolff, the CIA's team of brainwashing researchers, were equally keen to 'dispel the mystery of brainwashing'.[99] Although they did not mention him by name, referring to him instead as 'a reporter who was interviewing Chinese refugees', it is clear that they blamed Hunter for the anxiety among the public about 'special oriental knowledge of drugs, hypnosis, and other exotic and devious means of controlling human behavior'.[100] It was the public's fear, together with the embarrassing public confessions of American servicemen, that was the most worrying consequence of brainwashing. But there was nothing exotic about brainwashing; it could be understood in terms of 'ordinary psychological principles'.[101]

An important aspect of the ordinariness of 'brainwashing' according to these academic researchers was that it was not based on science. These techniques had not been developed by scientists, nor were they administered or supervised by scientists. 'The methods employed appear to be rather ordinary techniques of coercion and persuasion such as laymen have adopted at various times in many lands.'[102] Their origin lay in old interrogation methods such as those employed by the Inquisition, by witch hunters and by other repressive forces. The Soviet secret police had developed them further in the interbellum but by trial and error, not with the help of scientists. In fact, Hinkle and Wolff noted, Russian interrogators held psychiatry and psychology in contempt, and they considered the polygraph a useless gadget.[103] Neither Pavlov nor his followers had been involved in the development of 'brainwashing'. Raymond Bauer[104] thought it 'particularly disturbing to find a number of otherwise responsible citizens accepting the thesis that "brainwashing" is based on psychiatric principles developed from Pavlovian psychology'.[105] Based on his extensive knowledge of Soviet psychology he could confidently say that there was no indication that Russian psychologists or other scientists had contributed to the development of 'brainwashing'. Hinkle and Wolff concluded that we should not dignify these methods 'by surrounding them with an aura of scientific mystery, or to denote them by terms such as "menticide" or "brain washing" which imply that they are scientifically organized techniques of predictable effectiveness'.[106]

Breaking Resistance, Shaping Compliance

Indeed, if one strips the texts of Hunter and Meerloo of the scientific mystery of hypnosis and conditioning and takes away the hyperbole of 'human insects' and 'robots', one is left with the rough outlines of 'brainwashing' as Biderman, Hinkle and Wolff and their colleagues saw it too. These researchers presented the same facts about the Russian and Chinese interrogation and reform practices but without the insects, the robots and the insistent tone of alarm, and with different explanations of how they worked. Like Hunter and Meerloo, they all concluded that the Russian and Chinese practices that had become known as 'brainwashing' consisted of two parts, which to some extent corresponded to two phases. First, the prisoner's resistance was broken, and then his compliance was shaped into the correct behaviour. Hinkle and Wolff summarised it as 'demoralizing' and 're-organizing'.[107] Robert Lifton called it 'an agonizing drama of death and rebirth'.[108]

Russian and Chinese interrogators used a variety of methods to 'soften up' their victims, all of them unpleasant, but physical violence was seldom part of them. The Russians put their prisoners in solitary confinement immediately after their arrest.[109] The light stayed on day and night; the food was barely sufficient and of poor quality. The prisoner was allowed to sleep in only one position. If he moved, the guard woke him up. Interrogations took place at unpredictable moments but always at night. The interrogator waited until the prisoner had just fallen asleep and then had him woken up and taken to the interrogation room. If necessary, extra pressure was put upon the prisoner by making him stand in a fixed position for long periods of time, leading to swelling of the legs and severe pain. Within six weeks, prisoners were thoroughly confused and demoralised, their critical and judging capacities severely affected.

The Chinese also used sleep deprivation and bad food to break the resistance of their prisoners but not solitary confinement.[110] On the contrary: prisoners were housed in crowded cells, so crowded, Lifton writes, that 'during the night, prisoners cannot turn individually, but must do so en masse at a signal from the cell chief'.[111] Visits to the toilet were restricted to twice a day, when the six to eight cellmates had to run to a few open toilets where they were given little more than half a minute to do their business. New prisoners were shackled with their hands behind their backs so that they had to eat their meagre rations like a dog from its bowl. They were housed with cellmates who were further advanced in their re-education process and who could improve their own chances of release by doing their bit to reform the new 'student'. When he returned from an interrogation, the others crowded around him and began a 'struggle' session: they hurled insults at him (the Chinese seemed particularly intent on humiliating their prisoners), accused him of being an imperialist and a reactionary and told him to be sincere and confess his terrible crimes. These 'struggle' sessions could last all day.

Usually, neither the Russian nor the Chinese interrogators would actually inform their prisoners exactly what they were being accused of, instead telling them repeatedly that they knew very well what their crime was. In most cases, a prisoner would eventually reach a point where he would be willing to meet his interrogator's demands, whatever they were, if that would bring relief from the harsh treatment. Now the second phase of the process started. Suddenly his captors became friendly, even apologetic. There were tea and cigarettes on the table. If only he would confess, the prisoner was told, he could be out of here in no time. Deeply grateful for the sudden leniency, the prisoner now tried to find out how to please his interrogators. He 'confessed' something that he thought his captors might see as a crime. Although they rewarded him lavishly for it, it was not enough. He had to tell more. If they were not satisfied, the tea and cigarettes were taken away and he was put back in the old, harsh regime. Feverishly the prisoner continued to extend and revise his confession, tailoring it to what he thought were the wishes of his captors. It wasn't easy – Biderman noted that what appeared to be heroic resistance in some cases was simply due to a prisoner's inability to figure out what the interrogator wanted.[112]

It was not enough to produce a satisfactory confession or the appearance of a conversion. Russian and Chinese interrogators and re-educators insisted that the confession or conversion be sincere, that the prisoner should really believe what he was saying. It was this final phase, making the prisoner believe his own statements, that was 'the very essence of brainwashing', according to Hinkle and Wolff.[113] Essentially the prisoner now had to teach himself the truth of the fiction that he had produced. The prisoner was forced to retell or write out his confession time and again. Russian investigators would point out inconsistencies in the text, forcing the prisoner to defend his own fabrication. Terrified of going back to solitary confinement, he passionately argued that this was really the way it happened. The Chinese were even more fussy about sincerity. In the 'struggle', glib confessions or facile statements of allegiance to

Communism were mercilessly attacked as 'window dressing', 'spreading a smoke screen' or 'failing to combine theory with practice'.[114] Only a heartfelt conversion to 'the people's standpoint' was enough to bring about release.[115] Chinese thought reform could take years.

The Mind of the Victim

'Brainwashing' was a process of creating compliance and then moulding it into the correct form, and it could be explained by what Hinkle and Wolff called 'ordinary psychological principles'. On this the academic researchers agreed. Opinions differed, however, about what these principles were. A wide variety of psychodynamic and social-psychological concepts and theories were proposed to explain what happened between interrogator and victim and what went on in the mind of the victim as a result. Some researchers, but not all, emphasised the role of guilt in brainwashing, just like Meerloo had. Robert Lifton, for example, thought that in Chinese brainwashing the relentless condemnations and accusations become merged with the prisoner's own feelings of sinfulness, creating a powerful 'guilt anxiety'.[116] Desperate for atonement, the prisoner would become motivated to help his captors. Schein developed a model of brainwashing, or 'coercive persuasion' along the lines of Lewin's field theory.[117] In Chinese thought reform a person's beliefs, attitudes and behaviour patterns are changed by manipulating the various forces that act on a person, such that his existing equilibrium is unfrozen, changed to a new state and refrozen. Like Lifton and Meerloo, Schein also gave guilt an important role in the process because the Chinese required that a confession be based on a true feeling of guilt.

The work of Hinkle and Wolff, who produced four versions of their report in quick succession, encapsulates the variety in theories of brainwashing.[118] In the first version, they described the state of mind of the prisoner as dominated by the conflict over whether to give in to the demands of the interrogator. The 'very core of brainwashing', they wrote, is the prisoner's own resistance and the internal conflict that results from it.[119] As the hardship continued and the interrogator increased the pressure, the conflict intensified because the prisoner realised that it was 'his own original determination to resist that is causing the continuance of pain'.[120] Gradually, the interrogator would become the central figure in the prisoner's internal conflict. He was 'the protagonist around whom the prisoner develops his conflict, and upon whom the prisoner comes to depend as he seeks a solution of that conflict'.[121]

In the second version of their report, Hinkle and Wolff made 'frustration' the central concept. The sleep deprivation, the isolation, the forced positions, they were all intensely frustrating experiences that brought the victim to a state of 'emotional bankruptcy' in which he was ready to accept or do anything that promised relief.[122] In this respect the victims of Communist interrogation techniques were rather like psychiatric patients, whose problems frustrate them in the same way until they are so emotionally drained that they are finally ready to accept treatment. The victim of Communist control techniques was an 'intentionally created patient'.[123]

The analogy between psychotherapy and Communist interrogation returned in the third version of the report, an article published in the American Medical Association's *Archives of Neurology and Psychiatry*. Although they mostly avoided theory, on the grounds that 'many present-day concepts of human behavior are still in a formulative [sic] state', they did propose that it is 'helpful to consider the individual man as a living system entirely dependent upon maintaining a satisfactory relationship with his total environment'.[124] The frustrations of the Communist imprisonment and interrogation techniques disturbed this relationship, thus turning the prisoner into an artificial mental patient.

A few years later, Hinkle expressed his exasperation with all the theorising about brainwashing. In a book review he scorned Lifton and Schein for the 'excessively high' ratio of explanation to information in their books.[125] Lifton's book at least was readable, although he was more confident in his theories than seemed warranted, but Schein's theoretical discussions were merely dull and pedantic. What was needed was more observations, not more explanations.

Personal Relations

Notwithstanding the diversity of theories, all researchers agreed that the relationship between the interrogator and his victim was essential in the 'brainwashing' process. Hinkle and Wolff noted that the interrogator would try to create 'a bond of companionship' between him and the prisoner.[126] Schein reported that the Chinese thought reform teachers in the Korean POW camps often tried to develop a close tie with individual 'students'. Since they had been separated from their friends and their unit, many POWs were receptive to such attempts. Intense personal relationships could develop, and sometimes the interrogator even 'lived with his subject and tried to create an atmosphere of warmth and friendliness'.[127] Such an individual approach was more successful than the lectures and democratic group discussions that formed the bulk of the programme, which gave the POWs ample opportunity for evasion and insubordination, particularly because few of the Chinese were skilled enough in English to stay in control in the discussions.

To be most effective, the interrogation techniques had to be adapted to the individual victim. Russian and Chinese interrogators were flexible and tailored their approach to each specific individual. The interrogator had to have the 'ability to diagnose what combination of pressures will be most effective in manipulating a particular personality' and plan a treatment accordingly.[128] Although Chinese thought reform was mostly a group process, Chinese interrogators nonetheless needed to have insight into the individual prisoner to move him towards a confession, which was facilitated by having the victim write an autobiography, keep a diary and engage in public self-criticism.

Lifton came closest to attributing what Hinkle and Wolff had called 'predictable effectiveness' to the methods of the Communists. Although there was nothing mysterious about the procedures of thought reform – they were well-known 'religious, therapeutic, and hypnotic methods' – he thought they were novel in one important respect: all the elements were bound together into a tight, cohesive procedure by the unparalleled control over the environment.[129] It was this overwhelming 'milieu control' that made thought reform a mechanical, repeatable process regardless of the personality or cultural background of the prisoner.[130] The all-powerful external milieu simply replaced the prisoner's internal milieu, turning him into a helpless puppet of his captors. The free world better beware, Lifton warned: 'In a world of great technological and psychological advance, milieu control is a method which bears watching.'[131] It may well be 'the most advanced method of "soul engineering" yet devised'.[132]

But although Lifton stressed the mechanical efficiency of Chinese thought reform, it was not entirely machine-like, nor was it fully effective. Firstly, thought reform still required a personalised approach. In the re-education phase, each prisoner had his own instructor, who used his knowledge of the life history and psychological weaknesses of a prisoner to undermine his world view and thought patterns. Prisoners referred to the instructor as 'the analyst', and each one developed an ambivalent, complex relationship with his analyst.[133] Like Hinkle and Wolff, Lifton saw similarities with psychotherapy in thought reform – part of the re-education phase, for example, was the 'working through' of the newly acquired insights.[134] Secondly, thought reform was not 100 per cent effective or fool proof. The human being is remarkably resilient, Lifton noted, able to mount resistance when under severe stress. And in most cases, as soon as the prisoner was released, his mind started to revert to its former state. Gradually he began to understand what had happened to him and to question his new beliefs. Most former victims regained their previous ideas.[135]

Coercion and Subtlety

For Hunter and Meerloo, the evil essence of brainwashing lay in its technological side, which they both equated with Pavlovian conditioning. They stressed the impersonal, inhuman character of brainwashing and menticide: these methods worked by mechanical repetition, incapacitating consciousness by their hypnotic effect, conditioning the subconscious machinery of the brain into new beliefs and behaviours, turning people from conscious, free and unique individuals into identical puppets, robots controlled and manipulated by the state. Brainwashing and menticide were technological in their operation and in their result. But although they put the emphasis on hypnosis and conditioning, Hunter and Meerloo had also noted the personal aspects of the interrogation and reform practices: both saw similarities between brainwashing and psychotherapy and between the psychiatric patient and the victim of brainwashing. Both had pointed out the prominent role of persuasion and argumentation and the connection with everyday processes of influence. This was in essence how the academic researchers saw 'brainwashing': a process that was personal in character, one that was based on what Hinkle and Wolff called 'ordinary psychological principles'.

If one ignores Hunter's and Meerloo's ideas about hypnosis, conditioning, and human robots, the Communist methods appear more as a matter of tact than of technology. They required the medium of a personal bond between interrogator and prisoner, and only worked if the interrogator was able to adapt his actions to the particular individual before him and to the state he was in at that particular moment. The interrogator or teacher had to create a relationship with the prisoner or student and become familiar with his personal history. He had to figure out the emotional weaknesses of the individual prisoner and exploit them, to sense when he was close to breaking and it was time to switch from pressure to leniency. Control was essential, of course, but so was the ability to improvise and work with the opportunities that presented themselves. Schein's term 'coercive persuasion' for these interrogation and reform practices neatly summarises the views of all these researchers.[136] These practices were a combination of brutality and tact, torture and rhetoric: the deployment of subtlety and sensitivity in a situation that the interrogator controlled with an iron fist. Coercion was obviously a crucial element in the process, but as in all forms of persuasion, the interrogator's sense of *kairos* was key to success.[137]

Even the power of the interrogator was relative. According to Melley, 'the theory of brainwashing studiously avoids structuralism; it preserves the intentionality at the heart of individualism by understanding social control as the work of an exceptionally powerful, wilful, rational, and malevolent human agent – the brainwasher'.[138] But although that was, and is, the popular image of brainwashing, in many of the studies, including those of Hunter and Meerloo, the power of the system over the individuals is emphasised. The whole point of Meerloo's *Rape of the Mind* is that the power of the rational individual has become compromised in modern society. The real villain in the book is not so much the brainwasher or even the party; it is the technologisation of society that has allowed processes of influence to grow rampant, unchecked by individual rationality and morality. All social interaction is shot through with mutual influence, which is always, moreover, potentially coercive. Our infatuation with technology is robbing us of the moral and intellectual strength to rise above influence and control it with reason. We're delivering ourselves to the totalitarian mechanism of technology.

Other authors also relativised the power of the brainwasher, albeit in less grandiloquent prose. Hunter noted that Chinese interrogators were under immense pressure to produce results. If they didn't meet their quota, they were severely penalised. As a result, the interrogator is often as anxious as the prisoner; '[f]ear permeates both sides in the communist confession ritual'.[139] Hinkle and Wolff described a similar dynamic in Soviet interrogations. In one of the public versions of their report, Hinkle and Wolff explained the peculiar need to have prisoners make confessions even when they are known to be false. Soviet law stated that people may only be arrested if there is reasonable suspicion of a crime. Hence, when the KGB arrested someone at the behest of state officials because he was considered to be a threat, it was up to the KGB officer to produce a confession of a crime so that the judge would consider the arrest to have been lawful. Otherwise the officer would be censured. As a result, 'any man who is arrested is automatically in the position of being "guilty"'.[140] Thus, although the interrogator controlled the situation of the prisoner, he was himself stuck in an all-enveloping system that determined what he should achieve. '[T]he interrogator must follow the system, and there is nothing which he can do about it.'[141] Some prisoners even confessed because they took pity on their interrogator when they realised what a difficult situation he was in.

Secondly, the power of the interrogator was limited by the very aim of the whole procedure: to get the prisoner to take responsibility for the new 'truths' that he was supposed to profess from then on. They were to be *his* sincerely held beliefs. In the end, the prisoner had to become the main actor, not the interrogator. Thus, once the harsh regimen of imprisonment had motivated the victim to find a way out of his misery, the skill of the interrogator lay in offering a rationalisation that made his confession or conversion look like the prisoner's own choice, corresponding to his own standards, and not like a capitulation. The interrogator, armed with the knowledge about the beliefs and values of the prisoner that he had amassed during the interrogations, used any overlap with Communist dogma, however small, as a starting point to build a rationalisation.[142] Once the prisoner had accepted the 'way out', a decision that usually brought a huge sense of relief to both victim and interrogator, the process of constructing a confession or converting to the new beliefs was a collaborative effort. Sheer repetition was a part of it – each time the prisoner wrote down his 'confession' it became his own a bit more – but, just as in psychotherapy, the active participation of the prisoner was crucial. He wasn't simply being conditioned, and in as much as hypnosis came into it, it was largely as self-hypnosis.

If the brainwasher was not an all-powerful manipulator, perhaps it was better to focus on the system rather than the individuals. As noted, Lifton saw the Chinese prison as 'the most thoroughly controlled and manipulated group environment that has ever existed', but he located that control not in the interrogator but in the 'milieu', the 'system of communication – from officials to cell chief to cellmates to prisoner'.[143] Milieu control, he added, was never absolute but always limited by the remarkable capacity for resistance of human beings. The systems perspective, which had gained prominence in communications theory and cybernetics, was also applicable to the relationship between interrogator and prisoner. Thought reform, Lifton wrote, consisted of 'transactions', which were 'the interpenetrating and mutual ... reverberating effects of processes which can no longer be referred to organized, detachable entities ... a matter of system in process with system'.[144] To focus on the 'detachable entities' of interrogator and victim was to miss the processes at the level of the system that make thought reform what it is.[145]

All in all, as Hinkle and Wolff put it, '[t]he most effective features of the Communist procedures are those which would operate even in the absence of control'.[146] That is: the persuasive, subtle and tactful features. 'Brainwashing' was not something that was simply done to a victim by a powerful, controlling interrogator. It was a much more complex process, at the heart of which were two, related paradoxes: that a truth had to be fabricated and that the victim had to be forced to take responsibility for it. Control and coercion were essential to the conditions within which it took place, but they did not belong to the interrogator. They originated with the system that enveloped the prison, in which the interrogator to some extent was a mere pawn himself. Neither was coercion enough to turn fiction into truth and the victim into its proclaimer. The resolution of the paradoxes in the process depended on the interrogator's skill in finding out and exploiting the weaknesses of the victim, showing him a way out and persuading him to work together in constructing a confession or conversion that was acceptable to all given the circumstances. 'Brainwashing' was not a mechanical process, a technology of 'predictable effectiveness'. Rather it was a series of subtle, interpersonal transactions, every bit as complex as any rhetorical process.

Better Brainwashing

The unanimous conclusion that the Communist techniques of interrogation and thought reform were not based on scientific research allowed an inference that only Hinkle and Wolff put in writing. '[C]onsidering what is known of the brainwashing process', they wrote in the first version of their report for the CIA, 'more systematic application of established psychological principles could probably increase the efficiency of brainwashing'.[147] Not surprisingly, they didn't mention this possibility in the public versions of their report. There was, in fact, no need to point it out to the CIA because the agency had been conducting its own research into mind control ever since its establishment. By the time that Hinkle and Wolff and the other researchers reported that 'brainwashing' was not a scientific technique of predictable effectiveness, the CIA had already been chasing the dream of control over human behaviour for nearly ten years. It would continue to do so for some time with the help of, among others, Hinkle and Wolff.

The CIA, created in 1947, was interested in mind control right from the start, driven by what John Marks has called '[t]he ancient desire to control enemies through magical spells and potions'.[148] During the Second World War, the idea that drugs, hypnosis or a combination of the two could be used to control a man's behaviour and, for example, turn him into an assassin had been mooted already in American intelligence circles.[149] Mindszenty's strange behaviour during his trial seemed to indicate that the Russians had in fact discovered an effective thought control technique and that a 'mind control gap' had opened between the Communists and the free world.[150]

Irving Janis, psychologist at Yale University, provided the immediate spur to set up a mind control programme. In a research memorandum for the Air Force's Project RAND, Janis explored the hypothesis that false confessions like Mindszenty's were the result of the application of hypnosis, possibly in combination with drugs.[151] Janis quoted results from studies by the Russian psychologist A. R. Luria, who, by hypnotic suggestion, had succeeded in implanting false memories of misbehaviour in subjects with the associated feelings of guilt. This and other, similar findings suggested that it was at least possible that the false confessions in show trials were the result of hypnotic techniques.

Janis's faith in the claims of hypnotists[152] was enough to make him assure his readers that someone could be hypnotically trained to make a detailed confession at a predetermined signal and no one would notice anything unnatural about it. The false confessions, in other words, were essentially the result of remote control.[153] Ideas were implanted in the victim by hypnotic suggestion and then elicited from him in the courtroom by a trigger that he was hypnotically trained to respond to. The inevitable initial resistance to hypnosis was probably overcome by administering drugs or by hypnotising the victim in his sleep. Electroconvulsive treatment could be used to make people more hypnotisable. Clearly, the sensational ideas about the use of hypnosis in brainwashing that would later circulate had an academic pedigree.

Janis recommended that 'the general hypothesis that hypnotic techniques are being used by Cominform police agencies to elicit public confessions' should be further studied and, secondly, that experimental research should be started to 'duplicate' the Communist achievements.[154] 'E.g., can we take a group of confirmed atheists and, by means of specially devised hypnotic techniques, get all of them to accuse themselves of past iniquities in the sight of the Lord, and to confess to a number of sins they had never actually committed?'[155] Janis finished by saying that the moral and political implications of such research should also be explored, given that it lent itself to offensive as well as defensive applications and to abuse as well as use. Of these three lines of study, the CIA, predictably perhaps, chose first to pursue the goal of duplicating the Russian results to close the mind control gap as soon as possible. Only later, at the height of the brainwashing scare that it had itself helped to create, did the CIA put Hinkle and Wolff to work to find out more about what the Communists were actually doing. The third subject had no priority: the history of the CIA's mind control research shows few signs of a concern with moral issues.

Mind control studies were administratively organised in four consecutive CIA projects, named BLUEBIRD, ARTICHOKE, MKULTRA and MKSEARCH, all top-secret.[156] Of MKULTRA, the longest project, only the financial files remain; the rest were destroyed at the behest of CIA director Richard Helms when he was removed from office. The remaining archive of the four projects, reduced as it is, nevertheless shows clearly how doggedly, desperately even, the CIA clung to the dream of a scientific, predictably effective set of techniques for the control of mind and behaviour that could turn spycraft from an art into a technology. As one researcher told John Marks: 'The operators would come to us and ask for the human element to be removed.'[157] Yet what the many research projects of BLUEBIRD, ARTICHOKE, MKULTRA and MKSEARCH ended up proving was that effectiveness could be achieved only by taking the human factor into account. The CIA-funded brain and behaviour scientists were instrumental, sometimes in spite of their own aims and dreams, in dashing the hope for a magic bullet. Two examples must suffice as illustration.

Hypnosis

Early on, drugs and hypnosis were the two main contenders for a miracle mind control tool. Morse Allen, who led both BLUEBIRD and ARTICHOKE, had a personal fascination with hypnosis. He took hypnosis training, read everything he could find about the subject and experimented on CIA personnel, all of which seemed to indicate that hypnosis had great potential, if necessary in combination with 'sleep induction' via sodium amytal or sodium pentathol, just as Janis had advised.

One BLUEBIRD memorandum described four research aims: to ascertain the value of hypnosis for obtaining information from willing and unwilling individuals, for conditioning CIA personnel or others to prevent outside powers from obtaining information from them, for gaining 'control of the future activities (physical and mental) of any given individual' (the remote control paradigm again) and finally for preventing such control by outside powers.[158] The BLUEBIRD team could already answer the first question in the affirmative and was positive that further research would make the other three forms of control possible as well. The team had found someone with considerable experience in sleep induction and hypnosis and proposed to employ him for a series of studies with an emphasis on controlling subjects via post-hypnotic suggestion. 'Could we seize a subject and in the space of an hour or two by post-H control have him crash an airplane, wreck a train, etc.?'[159]

The hypnotists that Morse Allen consulted were eager to confirm his optimistic expectations. One of them, 'a professional hypnotist of exceptionally good reputation',[160] described a secure way to transmit messages by courier: 'As you know, an individual who has been hypnotized makes a very excellent courier. They can be given messages when under hypnosis which they themselves do not know and only an individual knowing the code can get this message from this courier.'[161] To further secure the message the courier could be hypnotically 'regressed' to a certain time and date and then given the message; the receiver would have to regress the person to the correct time and date to get the message out.[162]

This hypnotist also made it clear, however, that hypnosis is not a simple trick. Not only does it require great skill, it must always be adapted to the particular individual being hypnotised. 'All individuals are different and a hypnotist always studies his subjects carefully, psychologically and otherwise, before attempting hypnosis with them.'[163] Likewise, the idea of a simple pharmacological solution to the challenges of intelligence work should be abandoned. 'I do not believe there is anything such as the "wonder" drugs which will produce instant talking when given to an individual.'[164] To Allen's chagrin, academic hypnosis experts were even more reserved. One thought it was morally wrong to use hypnosis for such purposes but found Allen's questions nonetheless intellectually interesting.[165] He felt there 'was a positive use for hypnotism' but warned that the information so extracted is not always reliable, 'since fantasy and even hallucinations are present in certain hypnotic states'.[166]

Despite such warnings, the search for a simple technology of control continued. Allen's hypnosis studies had been almost endearingly amateuristic, but soon the CIA was funding the work of a whole host of legitimate behavioural, social and medical scientists, many of whom had no idea that their funding came from the CIA. The money was distributed through conduits like the Society for the Investigation of Human Ecology, founded by Harold Wolff and Lawrence Hinkle. The society, 'the Agency's window on the world of behavioral research', was a major influence on the behavioural sciences in the late 1950s and early 1960s.[167] It was willing to finance unconventional research and had ample funds to do so.

Adventures with Repetition

The most infamous research project funded through the Society for the Investigation of Human Ecology is also the one that is most characteristic of the CIA's control fantasies, and it eventually showed those fantasies to be misguided. Dr. Ewen Cameron was a celebrated psychiatrist, president of the American Psychiatric Association and head of the Allan Memorial Institute of Psychiatry in Montreal.[168] Cameron believed psychotherapy could be automatised to the benefit of all concerned, not least the patient. To this end he had devised a technology that he called 'psychic driving'. When one of the CIA's staff psychologists, trawling through the literature for promising new findings and techniques, came across Cameron's description of his new treatment method, he immediately recognised 'psychic driving' as a mechanised form of brainwashing. There is some controversy over whether Cameron was aware of the origin of the funds, but in 1957 the CIA began to sponsor his research via the Human Ecology front. His work became MKULTRA subproject 68.

Psychic driving consisted of repeating to the patient, via a tape recorder or phonograph, statements of himself ('autopsychic driving') or the therapist ('heteropsychic driving').[169] Autopsychic driving was particularly useful for penetrating a patient's defences: repeating a carefully chosen statement for up to thirty minutes was usually enough to make the patient realise the meaning of what he or she was saying. Next, heteropsychic driving could be used to change a patient's attitude or to set up a 'dynamic implant', a long-lasting activation of a 'community of action tendencies' that forced the patient to reflect on them and reorganise that particular mental area.[170] Heteropsychic driving could continue for a long time: 'Eventually our patients were listening to verbal signals ... from 6 in the morning until 9 at night day after day and week after week.'[171]

Driving was so effective because it left the patients no escape from facing their issues. If necessary, Cameron would drive his patients in their sleep, using speakers in their pillow or on the ceiling, or he would 'prepare' them with LSD, stimulants or sleep-inducing drugs, the latter allowing him to continue driving for up to twenty hours a day for fifteen days.[172] Sensory deprivation was also a potent way to focus patients' attention. Unable to ignore the 'endless repetition of the cue verbalization', patients would enter a self-reinforcing process of the continuous reactivation of a 'community of concepts'.[173] Using an analogy that should have warned his readers about what Cameron was doing to his patients, he described it as 'a situation comparable to the fire blizzard of World War II – when the hotter the conflagration grew under the rain of incendiary bombs, the more air poured in, and hence the more intense grew the conflagration'.[174] Psychic driving is also analogous, he added, to 'the breakdown of the individual under continuous interrogation'.[175] The connection between psychic driving and brainwashing that the CIA made had not been lost on Cameron. In fact, Leonard Rubenstein, Cameron's technical wizard, later claimed that psychic driving was inspired by the reports about the treatment of POWs in Korea. 'It was directly related to brain-washing.'[176] A 1955 magazine article touted Cameron's method, presumably with his permission, as 'beneficial brainwashing'.[177]

Psychic driving automated the treatment of neuroses, but Cameron also sought a more mechanical approach to psychoses. To this end he developed an even more extreme therapeutic technique, which he called 'depatterning'.[178] Cameron tried to treat paranoid schizophrenia with a combination of prolonged sleep and electroshocks. Patients were kept asleep with a cocktail of drugs (chlorpromazine and three barbiturates) and only woken up a few times a day for meals and toilet breaks. After ten days of sleep, electroshocks would be added to the regimen. Cameron found the Page-Russell technique quite effective: it consisted of a series of six shocks in quick succession. Patients received two Page-Russells a day.[179] Soon, their mind would start to depattern: first their memory would start to falter, next they would forget where they were and why, and finally they could not recognise anyone anymore, but this did not trouble them. '[T]he patient is now quite smiling and unconcerned.'[180]

Once this depatterned state had been maintained for up to a week, the rehabilitation phase began. Cameron no longer considered depth psychology of any use for bringing his patients back to sanity. Cameron's theory was that depatterning would produce 'differential amnesia': it removed the schizophrenia, but the rest would spontaneously return in a suitably supportive environment.[181] Thus, patients were simply discharged and returned to the hospital once a month for electroshock treatment. In case of 'turbulence', a return to confusion, the patient received 'heavy doses of Thorazine, preferably intramuscularly, to control the patient's anxiety for the next three to seven days'.[182] Cameron also realised that he could combine the two techniques: psychic driving after depatterning. It was this combination that he originally received funding for from the CIA.

Psychic driving and depatterning did not become the revolution in psychiatry that Cameron had hoped for. None of his colleagues outside his own hospital adopted his methods, and over the years the resistance in his own institution grew as well. In practice, the success rate of Cameron's treatments was a lot lower than he claimed in print – Cameron preferred to report only the successful cases.[183] Depatterning was certainly predictably effective, but rehabilitation did not always follow, nor did psychic driving usually succeed in repatterning a patient's mind. Many patients were seriously harmed, particularly by depatterning, suffering lasting mental and physical problems. Some patients never recovered their memories of their lives before they were admitted to the Allan Memorial Hospital. By 1960 the CIA had also become convinced that Cameron's work offered little of practical value and terminated his funding.[184] It is not clear exactly why, but one can speculate that the CIA considered wiping someone's mind and reducing him to a state of oblivious contentment of limited use if one could not reorganise his mind in a specific way afterwards. There are easier ways to silence people.

Both psychic driving and depatterning only automatised psychotherapy in a superficial sense. The techniques involved material devices – microphones and tape recorders, a range of pharmaceuticals, a room specially equipped for sensory deprivation – and the interaction of ordinary psychotherapy was as much as possible abstracted into the repetition of selected phrases or simply abolished altogether. But the choice of statements to be repeated could not be automatised and depended very much on Cameron's clinical intuition. As he developed his techniques further, Cameron increasingly began to acknowledge the limits of automation. He noted that in the rehabilitation phase, after depatterning, 'building a strong relationship between the patient and the therapist' and giving emotional and practical support were crucial.[185] Rehabilitation depended, moreover, on friends and family offering a 'scaffolding' by answering the patient's questions and helping him remember his past life. 'There is some danger', he wrote, 'of falling into a belief that since treatment as here described is largely by means of physical and chemical agents . . ., that the process is mechanical. Nothing could be further from the facts.'[186]

When Cameron looked back upon his 'adventures with repetition' in his 1963 presidential address at the annual conference of the American Psychopathological Association, he had to concede that setting up 'truly compelling patterns of behavior' had proved to be more difficult than he had imagined.[187] He did feel confident that the journey towards that goal would remain stimulating and rewarding, not least for the patients themselves.

The KUBARK Interrogation Manual

In the latter half of the 1950s the CIA gradually began to shift its focus, from developing a foolproof technique for turning people into puppets towards 'the reduced goal of influencing and manipulating their subjects'.[188] The experiments with hypnosis, electroshock, sensory deprivation, psychic driving, LSD and 'truth drugs' had not yielded a straightforward technology of mind control: there was no device, substance or procedure that by itself enabled the manipulation of another person. The studies of Russian and Chinese interrogation and thought reform practices made clear that physical control over the victim is only half the story. In order to make a person say or do what you want it is certainly important to control his physical environment as well as what he eats, how much sleep he gets and what information he receives. The ability to make him suffer pain when necessary and alleviate his suffering when opportune is crucial. But all of that is not enough. More subtle, tactful operations have to take place within that situation of control to make the victim say and do what is required.

These lessons were incorporated in the CIA's 1963 'KUBARK Counterintelligence Interrogation Manual'. The KUBARK[189] manual, declassified in 1996, represents what the CIA had learnt from seventeen years of research into mind control. It has become infamous as a 'torture manual' because among other things it discusses a series of techniques of coercion. McCoy has described KUBARK's methods as 'no-touch torture' – an indirect, psychological form of torture that the CIA had clearly adopted from the Communists.[190] But this was not the only lesson the CIA had drawn from its research. To focus solely on the coercive methods of the manual, as McCoy does, underestimates the complexity of its advice.[191]

An important source for the (unknown) compilers of the manual was the work of Albert Biderman. The extensive bibliography mentions six of his reports, articles and books. Like Hinkle and Wolff, Biderman had first tried to explain the mystery of the false confessions, but he had also turned his attention to a subject that was of more direct relevance to the CIA's own operations: how the Chinese succeeded in obtaining (correct) information from their prisoners. One of his reports for the Air Force, 'A Study for the Development of Improved Interrogation Techniques', is praised in the manual as 'one of the most useful works consulted'.[192] Unfortunately, the Pentagon has been unable or unwilling to release the report,[193] but it was likely based on a study that Biderman reported on at the 1958 annual meeting of the American Sociological Association.[194]

From the accounts of Air Force personnel taken prisoner in the Korean War Biderman concluded that violence is not effective in interrogations. What forced a prisoner to talk was not physical torture – in fact, the less coercion, the more compliance. What did work was the kind of 'self-inflicted pain' from standing at attention for extended periods and the 'serious debilitation' from lack of sleep, hunger and solitary confinement, what McCoy has called 'no-touch torture'.[195] However, building on the work of Erving Goffman, Biderman located the main source of compulsion in 'the requirements for the maintenance of the self and a viable, complementary role in social interaction'.[196] It is the interrogator's skilful manipulation of the dialogical situation that moves prisoners to talk despite their firm resolution not to. The norms and roles inherent in a conversation are so deeply ingrained that they are very difficult resist. Chinese interrogators would, for example, first ask the prisoner a series of questions that he could impossibly know the answer to. When the prisoner was finally asked a question he could answer, he would jump on the chance to settle the 'debt' he had unconsciously incurred. Or a prisoner would be tempted to talk with innocent questions about the weather or the food, and then the norm of consistency that is inherent in a conversation would impel him to keep talking, even in response to more pertinent questions.

Such conversation techniques are a central element of the KUBARK manual. It emphasises repeatedly that 'the interrogation situation is intensely inter-personal'.[197] By skilfully manipulating the forces that operate in that situation, the interrogator can coax the subject towards cooperation. Success depends, firstly, on establishing rapport with the 'interrogatee' (as he is called in the manual). The principles that the manual formulates in this regard are reminiscent of Dale Carnegie. Make sure to treat the subject as an individual; call him by his name when you enter the room. Speak his language, know his culture, and try to understand him 'as a person'.[198] Perhaps 'friendship', as Carnegie would call it, is too big a word for the recommended relationship between interrogator and interrogatee, but it is imperative to create genuine rapport. What makes the task of the interrogator especially difficult, however, is that he must operate on two levels: rapport must be combined with cool observation without the two interfering with each other. The interrogator must simultaneously be sincerely engaged and objectively detached. 'To be persuasive, the sympathy or anger must be genuine; but to be useful, it must not interfere with the deeper level of precise, unaffected observation.'[199]

As an interpersonal situation, each interrogation is unique. The interrogator must always adapt his tactics to the personality of the interrogatee, his strengths and weaknesses. Preliminary screening is therefore important, and while the interrogator is establishing rapport with a new subject, he must at the same time verify the information his colleagues have provided about the person in front of him. The manual lists nine 'psychological-emotional categories' of interrogatees that he may use as a guideline if he finds them useful.[200] He must, moreover, be aware of his own personality traits and how they may interact with those of the interrogatee. The uniqueness of the people involved makes an interrogation difficult to predict, but although it is impossible to plan an interrogation from A to Z, with careful preparation 'it can and must be planned from A to F or A to M'.[201]

In the KUBARK manual, coercion is a last resort. It urges the interrogator not to assume at the outset that the subject will resist. The first question that you ask yourself should always be 'How can I make him want to tell me what he knows?'[202] (Again, Dale Carnegie would have wholeheartedly approved.) If you start the process in a hostile manner, intent on breaking resistance, you will only create resentment. However, if a subject cannot be brought to compliance with conversational techniques and the like, sterner measures will be required. The manual describes the intended effect of coercion in psychodynamic terms: coercive techniques are essentially 'methods of inducing regression of the personality to whatever earlier and weaker level is required for the dissolution of resistance and the inculcation of dependence'.[203] They include sensory deprivation, threats, narcosis, pain and 'debility' through exertion, lack of sleep or food and extreme heat or cold.[204] But all these measures must be used with great circumspection. If coercion is not lifted in time, the subject may sink into complete apathy. It is imperative that the interrogator recognises a subject's breaking point so that he may proceed to guide him towards compliance by offering a rationalisation. The interrogatee will only cooperate if he can do so without losing face. Here, again, it is necessary to adapt the rationalisation to the personality and personal history of the interrogatee to make it convincing. Finally, once the interrogatee has begun to cooperate, it is important to make him take the final step: ideally he should 'replace his new sense of emptiness with new values', those of the interrogator, and become an ally.[205] That way, future trouble is prevented, and he may remain a useful asset.

Although the aim is different, and the manual does not name it as such, successful interrogation in the style of KUBARK is not all that different from Communist brainwashing as it had been described by Biderman, Hinkle and Wolff, and their colleagues: ruthlessly coercive when necessary but always tactful and adapted to the particular individual concerned and ultimately a cooperative endeavour, something you do together, albeit that the interrogator holds the monopoly on violence. As in Communist brainwashing, the interrogator, in turn, does not own this monopoly: the KUBARK manual stresses that the use of coercive measures is only allowed after permission from headquarters. Blanket permission is not given.

Suspicious Minds

Suspicion and anxiety permeated the discourse about brainwashing in the 1950s: anxiety about the mind control capabilities of the Communists, in the first place, and suspicions about their plans to subjugate the free world by brainwashing its citizens. But from the beginning this fear overflowed the boundaries of Communism and found forms of control closer to home to latch onto. Brainwashing could be found in many places. Wherever people changed their minds the question could be raised who or what had made them do that. Their insistence that they did so of their own accord could no longer be trusted. The false confessions in Russia, China and Korea opened up a frightening epistemological abyss: if it was possible to bring people to a sincerely held conviction and make them forget they had been coerced, then one couldn't even trust one's own beliefs. Control and manipulation could be anywhere. Meerloo drew the ultimate conclusion: the manipulator could be you. 'It is easier to manipulate the mind of others than to avoid doing so.'[206] It seemed that even control was out of control. Society was a web of influences and manipulation, some of it intentional, much of it unwitting and unconscious.

During the same period, concerns were raised about the methodology of the psychological experiment. Historian of psychology Jill Morawski has described the 'epistemological dizziness' that took hold of psychologists when they realised that their research subjects were not the passive recipients of stimuli that they were supposed to be, nor were they themselves the objective observers demanded by the handbooks.[207] Suspicion reigned in the laboratory: the subjects suspected they were being manipulated and resented the fact; the experimenters suspected that the subjects were manipulating and deceiving them in turn. Even in science people could not always be trusted, nor could scientists always trust themselves. Moreover, empirical studies of the social processes in the laboratory by Robert Rosenthal and Martin Orne, among others, uncovered mutual influences that went beyond the intended stimulus and the recorded reaction. The people populating the laboratory were much more complex beings than the official methodology assumed, and no one appeared to be fully in control.

In both cases – brainwashing and the psychological experiment – what started with anxiety about (someone else's) control developed into a generalised suspicion of both self and others and a view of the social situation as one that is shot through with processes of influence that no one seems to have in hand. But with regard to brainwashing as well as the psychological experiment the anxiety was eventually overcome with a more practical perspective. Both came to be seen as forms of interaction that were managed, not controlled. On closer inspection, brainwashing appeared to be a set of procedures and techniques that enabled the interrogator to handle the victim rather than fully control him. The victim never became a puppet – on the contrary, his cooperation was crucial – and success was not guaranteed, but the process was manageable. Similarly, psychologists pushed aside their epistemological angst by introducing a number of 'technical remedies' that did not give the experimenter full control over the situation but made it sufficiently stable and predictable to produce reliable results.[208] Experiments were acknowledged to be interactive rather than controlled, but as long as the interactions were scripted and the subject kept in the dark as much as possible about the aim of the experiment, the process could still yield truth. Of course there were fundamental differences between the situations: psychologists did not have to resort to torture to make their subjects compliant, nor did they usually engage in debate with them or try to get to know them personally.[209] But with regard to both kinds of situations – interrogation and experimentation – anxiety over total control was exchanged for a pragmatic attitude that aimed at managing social interaction.

But the dream of control did not die. Psychologists resumed the search for covert persuasion: ways to bypass the subject's consciousness and manipulate the hidden mechanisms of behaviour directly. The next chapter will feature one such technique. The intelligence community, for its part, continued to look for predictably efficient methods of controlling people. Currently much is expected of the neurosciences,[210] but the most significant recent development was the work of two psychologists.

Epilogue

Despite the nuances of the KUBARK manual, the CIA has continued to see torture as an efficient instrument to make prisoners tell the truth.[211] The latest tragic episode started a few years into the War on Terror.[212] Eager for a predictably effective method of interrogating terrorists, the CIA hired the services of two psychologists, Bruce Jessen and James Mitchell, who proposed a number of 'enhanced interrogation techniques' to speed up the process of obtaining information from resistant subjects. Once again, the CIA fell for the fantasy of mind control that it had itself helped to create. The two psychologists and the consulting company that they founded in 2005, Mitchell Jessen and Associates, received a total of $81 million for setting up and helping to carry out an interrogation programme that included the use of sleep deprivation, prolonged stress positions and waterboarding.

Jessen and Mitchell based their claims of efficiency and effectiveness on Martin Seligman's theory of learned helplessness, which, ironically, is in turn a descendant of Pavlov's work in conditioning. Seligman had discovered that dogs that had learnt that nothing they did could prevent the electric shocks they were given would remain apathetic long after the experiment had ended. Seligman theorised that depression is basically such learned helplessness. Jessen and Mitchell believed that the apathy of learned helplessness would make subjects more cooperative and willing to provide information. Repeatedly waterboarding a prisoner would eventually make him tell the truth.

The US Senate Intelligence Committee conducted a lengthy investigation of the CIA's detention and interrogation programme. Its summary, declassified in 2014, describes a number of 'enhanced' interrogations in graphic detail.[213] They appear to have consisted of little more than torture alternated with aggressive questioning.[214] The first of the report's twenty key findings was that '[t]he CIA's use of its enhanced interrogation techniques was not an effective means of acquiring intelligence or gaining cooperation from detainees'. No new and useful information ever came out of such interrogations.

9 The Priming Saga: The Subtle Technology of Psychological Experimentation

Introduction

In her analysis of the 'epistemological dizziness' that took hold of psychologists in the 1950s and 1960s, Jill Morawski notes that in such times of upheaval, the normally tacit, unnoticed preconditions of scientific practice are laid bare. Fundamental assumptions that structure the work of scientists, the social relations between them and the infrastructure that makes their work possible all become the topic of often heated discussion. In Morawski's words, '[t]he circuitry of the psychological is made more visible'.[1] Over the past five years, the machinery of psychology, the psychological experiment in particular, has again been subjected to inspection in a new period of upheaval that has been dubbed the 'crisis of confidence'. In this crisis, the issue of replication dominates the discussion. The question is precisely whether the experiment *is* a machine, a mechanism that can be expected to produce the same result time and again, regardless of who is operating it or where it is located. Much of the discussion, moreover, is focused on a field of research that studies the so-called 'automaticity' of human behaviour. It is a field that claims that, in important respects, human beings are machines that can be manipulated without their knowledge. Ironically, however, it appears to be difficult to stabilise the experiments that show this automaticity and reproduce their effects.

The current crisis shows that psychology still grapples with the relations between tact and technology, control and resistance, and conscious and unconscious sources of behaviour. The themes that I have discussed in earlier chapters have once again taken centre stage, but in contrast to the earlier episodes, the scene is now the laboratory. At issue is, among others, the replicability of the studies that underlie the applied work of people like Ap Dijksterhuis, mentioned in Chapter 1. Do the spectacular claims of subliminal, and thus irresistible, influences on behaviour hold water? But more than just particular experiments, the experiment as such has become the topic of intense discussion, and with it the question how to decide whether, for example, such subliminal influences are real.

Annus Horribilis

The year 2011 was a terrible one for psychology. Trouble had been brewing for some time, but in 2011 it erupted in an extraordinary series of scandals and heated arguments in scientific journals, in blogs and on Facebook and Twitter.[2] Hardly a week went by without upheaval, and soon a 'crisis' was proclaimed. Over time, the arguments have cooled down a bit, the pace of events has slowed, and two camps have emerged: a loose coalition of critics and a set of regular detractors, critics of the critics. The events of 2011 (between March 2011 and March 2012, to be precise) are still the reference points of the discussion.

The trouble started with an article titled 'Feeling the Future: Experimental Evidence for Anomalous Retroactive Influences on Cognition and Affect' by the American social psychologist Daryl Bem.[3] It appeared in the March 2011 issue of social psychology's number-two journal, the *Journal of Personality and Social Psychology*. The journal's ranking, measured by its so-called 'impact factor', is relevant, as will become clear below.[4] In the article, Bem claimed to have found evidence of precognition, hence the title. For example, in one of the experiments subjects were sat in front of a computer screen on which they saw two curtains, and they were told that behind one of the curtains was a picture. They were asked to pick the one 'that you feel has the picture behind it', and then the curtains were opened.[5] If the picture was of an explicit erotic nature, the hit rate was significantly above chance level. It was as if the subjects could sense where the rewarding picture would turn up. Nonerotic images did not have the same effect.

The article was followed by a commentary by a group of methodologists of the University of Amsterdam, which argued that the article was proof not of precognition but of the fact that '[t]he field of psychology currently uses methodological and statistical strategies that are too weak, too malleable, and offer far too many opportunities for researchers to befuddle themselves and their peers'.[6] The journal's editors added a brief statement about their reasons to publish Bem's piece, to the effect that they thought the reported effects were 'extremely puzzling', but the paper had gone through rigorous peer review and they trusted that the author had not cherry-picked the successful studies from a larger set.[7] In closing, they expressed the hope that 'the current two papers will stimulate further discussion, attempts at replication, and critical further thoughts about appropriate methods in research'.[8]

That, indeed, was a prescient remark. There did follow an intense discussion about methods of research and data analysis, and soon the issue of replication was also on the table. As it happened, an explicit goal of Bem's research programme was 'to develop well-controlled demonstrations of psi that could be replicated by independent investigators'.[9] A group of researchers duly attempted to reproduce the strongest of the results that Bem had reported and failed. They offered their report to the same journal, the *JPSP*, but it was rejected with the argument that the *JPSP* never publishes replication attempts, whatever their outcome.[10] Angry psychologists concluded that apparently it is fine to publish impossible results but not the studies that refute them. After one more rejection, the replication study was eventually published as 'Failing the Future'.[11]

The controversy about Bem's precognition study was still continuing when, right after the summer holidays, the discipline was rocked by a new scandal. Diederik Stapel used to be a well-regarded, upstanding and reasonably successful member of the social psychology community. When he was hired by my own university in 2001, he was considered a great talent, the 'golden boy' of Dutch social psychology. When he moved to Tilburg University five years later, he was given his own research institute. In September 2011 (by this time he was dean of the faculty of social and behavioural sciences) he confessed to having fabricated and falsified data. He was fired, and a commission was instated to investigate all his publications. At a stroke the problems in psychology had become international headline news, particularly when the preliminary report, published in late October 2011, showed that Stapel had been fabricating and falsifying data at least since 2004. Hadn't anyone noticed? Since the signs of malpractice in many of Stapel's publications were rather obvious in the commission's opinion, it concluded that there had been a 'failure of scientific criticism'.[12] Reactions to the intermediate report were mixed. The conclusions about the research culture in social psychology were grist to the mill of the growing number of critics, but others protested that Stapel's misdeeds could not be blamed on the field: he had been a rotten apple, but the field of psychology as a whole was healthy.

Rumours were already circulating, however, about two new cases of suspected fraud, both concerning social psychologists. They had been discovered through the detective work of Uri Simonsohn, a psychologist at the University of Pennsylvania. Simonsohn had discovered statistical anomalies in three articles and had requested the raw data of the studies.[13] Two of the three researchers sent him their data, and Simonsohn's suspicions of fraud were confirmed. He contacted their universities, which launched inquiries. Lawrence Sanna of the University of Michigan and Dirk Smeesters of the Erasmus University Rotterdam would later resign from their posts. The third case remains unresolved – the researcher claimed to have lost the data – and Simonsohn has chosen not to divulge the identity of the person concerned.

Still in October 2011, an article by Simonsohn and his colleagues Joe Simmons and Leif Nelson appeared online in which they exposed how easy the prevailing standards for reporting the details of one's study make it to claim evidence for false hypotheses.[14] Doing research, they noted, requires making many decisions: how big the sample should be, which variables to consider, how to analyse the data and many more. These are the 'researcher degrees of freedom', as Simmons et al. called them. Hypothesis testing requires such decisions to be made *before* the data are collected and analysed, a point that the Amsterdam methodologists had already emphasised in their reaction to Bem's precognition study. Simmons, Nelson and Simonsohn showed that by fully exploiting the researcher degrees of freedom *during and after* data collection and analysis, it was possible to produce evidence for a patently false hypothesis, in their case that people are on average nearly a year and a half younger after listening to the Beatles' 'When I'm Sixty-Four'. The laxity of journals' reporting standards allows researchers to hide these decisions and to pass off as a successful prediction what is in fact the post hoc discovery of a chance pattern in the data.

Like the Amsterdam methodologists earlier in the year, Simmons and his colleagues asserted that the tactics they described were 'accepted practice' as well as a source of error.[15] Empirical confirmation of this incendiary claim was forthcoming. Simmons et al. had seen a manuscript that was circulating of a paper about the prevalence of so-called 'questionable research practices'. The authors, Leslie John, George Loewenstein and Drazen Prelec, had surveyed more than 2000 psychologists.[16] Among the questionable practices they had inquired about were failing to report all of a study's dependent measures or conditions, deciding on whether to exclude data depending on how it influences the results and increasing the sample size until the results are statistically significant. That last practice, for example, was admitted by more than half the respondents. Although the self-admission rate of outright fraud was low, it appeared that many psychologists were wittingly or unwittingly employing methods that produce false positives.[17] Psychology seemed to have a bigger problem than just a few bad apples.

By this time, the word 'crisis' started to be used to characterise the state of the field. When, in January 2012, questions were raised about a famous study of a leading social psychologist, the sense of crisis became acute, particularly when the psychologist in question, John Bargh, reacted in such an intemperate fashion that he succeeded in drawing the attention of a wide audience in and outside the discipline. The targets of Bargh's ire were not only the researchers who had tried and failed to reproduce his classic experiment – they were 'incompetent' – but also the journal that published their article and an online science journalist who had blogged about the replication study.[18] The journal in question, *PLOS One*, is a so-called open access journal, meaning that its content is freely accessible but authors pay a fee for publication. Bargh hinted that this business model encourages open access journals like *PLOS One* to accept as many papers as possible, including weak papers such as this replication study. Without going into details, providing arguments or even mentioning his name, Bargh accused the science blogger (Ed Yong) of 'superficial online science journalism'.

Bargh published his attack on his blog 'The Natural Unconscious', hosted by the magazine *Psychology Today*. The comments section under his post quickly filled up with critical reactions: people pointed out that none of the supposedly essential elements of the experimental procedure, which the replicators had failed to include according to Bargh, were actually mentioned in the methods section of Bargh's original article. Others wrote to say that there had been earlier attempts to replicate that particular study that had also failed. The publisher of *PLOS One* wrote a comment to correct the factual errors in Bargh statements about his journal. Ed Yong pointed out that, when he was working on his own blog post, he had sent the replication paper to Bargh, asking him for his opinion of the study, but Bargh had answered that he didn't have much to say about it. Soon, supporters of Bargh also started to weigh in, arguing that the theory behind the work had been confirmed in many ways and complaining about the tone of the debate.

A few days later Yong reported on the affair in a new post, under the title 'A Failed Replication Draws a Scathing Personal Attack from a Psychology Professor'.[19] In the comments section of that post, as well as on other blogs, the discussion about Bargh, his experiment, the replication study, replication in general, open access publishing and many of the other subjects that had been so much talked about over the past year continued. In another ill-considered move, Bargh eventually removed his blog post, inviting accusations of a cover-up.[20]

Reproducibility and Transparency

A diagnosis of the problems in the field was beginning to take shape. Psychology suffered a reproducibility crisis, brought on by failings in the incentive structure of science. As it was formulated in a particularly influential paper on the crisis, 'getting it published' had become more important than 'getting it right'.[21] Under pressure to produce a high 'output' of articles in top journals (those with a high impact factor) and faced with the requirement of those top journals that articles present an eye-catching story backed by clean results, researchers often engage in questionable research practices (or worse) to manufacture just such results. As a consequence, journals are increasingly filled with spectacular but weak research. Bem's 'Feeling the Future' was a typical example. Because journals do not publish replications – see, again, the Bem case – researchers have very little incentive to check each other's work. Instead, they do so-called 'conceptual' replications, in which they build on each other's work by devising new experimental procedures to produce similar effects and extend existing theories. 'Direct' replications, in which the same experimental procedure is followed to see whether the original result is robust, are rare. Additionally, since negative results (experiments that 'do not work') are unpublishable, the field suffers from publication bias and an unwarranted confidence in its findings. The journals are full of successful experiments with interesting results, but these are a biased selection, with the boring failures languishing in the file drawer. The literature of the field appears to be a catalogue of spectacular successes, but it is in fact, in the words of two critics, 'a vast graveyard of undead theories'.[22] (It is customary at this point to add that psychology is not unique in suffering from such problems.)

In the solutions that were proposed, 'replication' and 'transparency' were the keywords. Unless (direct) replication studies became a standard practice, as the idea of science seemed to imply it should be, then the reliability of published results would remain doubtful and confidence would not return. To determine the extent of the problem, social psychologist Brian Nosek had started the Reproducibility Project.[23] Research groups from all over the world were taking upon themselves the task of replicating studies that had originally appeared in the 2008 volume of three psychological journals. The aim was 'to systematically estimate the reproducibility rate of psychological science as it is practiced currently, and to investigate factors that predict reproducibility'.[24]

Nosek was also taking steps to promote transparency. As it was, he noted, openness in science was realised almost entirely via research articles, but this limited transparency to the outcome of studies.[25] As had become painfully clear over the previous year and a half, that was no longer enough to inspire confidence in the robustness of the evidence. It was essential to open the entire research process for outside scrutiny: the data, the methods and tools with which they were collected, and the entire workflow of the study. To facilitate such 'open science', Nosek and Jeffrey Spies had set up the Open Science Framework (OSF, https://osf.io/), a web-based platform where researchers can house their projects and make them available for others to inspect, contribute to, or copy from. The Reproducibility Project, for example, is at https://osf.io/ezcuj/, where one can consult its individual replication studies, its overall analysis, all the data collected, the contributors and so on. The work on the OSF is now done by the Center for Open Science, also created by Nosek, which 'supports shifting incentives and practices to align more closely with scientific values' (https://cos.io/).

Amid the clamour of crisis talk, however, a few dissenting voices were also heard. Social psychologists Klaus Fiedler, Florian Kutzner and Joachim Krueger warned that the criticism had a 'cannabilistic aspect' and risked damaging the social impact of psychology.[26] Moreover, by exclusively focusing on false positives, results that do not correspond to true effects, the much larger and more fundamental problem of false negatives is both ignored and exacerbated. Stricter control of false positives will mean that true effects will go undetected and alternative hypotheses will be left unexplored. More precision at the methodological and statistical level is meaningless if the theories that are tested are not sound. But Fiedler et al. did encourage more replication studies: scientific progress is best served by weeding out false positives through critical replications, not by preventing them with statistical conservatism.

Yet the value of direct replication was not uncontested either. According to Wolfgang Stroebe, Tom Postmes and Russell Spears, direct replications have little value as a check on the validity of earlier results: 'In psychological research, there are always a multitude of potential causes for the failure to replicate a particular research finding.'[27] The description of the experimental procedure might have been insufficiently precise, or the replicating researcher might have lacked the required skills for this experiment. Perhaps there are variables involved in the effect that no one is aware of yet. Thus, a failed replication is not very informative. They were not the first, and would not be the last, to make this point. It has become a major bone of contention in the crisis, and it is directly relevant to the question of what the relation is between psychology and technology. The failed replication of John Bargh's study and the controversy around it illustrate perfectly how problematic this relation has become.

Priming and Automaticity

In the original experiment, a participant is told that she is taking part in a study of language proficiency. She is presented with thirty sets of five words, and her task is to make grammatically correct sentences using four of the five words of each set (a so-called 'scrambled sentences task').

When she has finished, the experimenter returns, and tells her that the experiment was about how people 'use words in various, flexible ways'.[28] He thanks the participant for her efforts and tells her the elevator is down the hall. On the way there, she passes someone sitting by the door of another office, apparently waiting for an appointment. When the participant has almost reached the elevator, the experimenter catches up with her. He asks her whether she had noticed that some of the words in the task were related to the 'elderly stereotype' and whether she thinks that had an influence on her.[29] The experiment, he tells her, was actually testing the hypothesis that activating the stereotype of the elderly person with words like 'wrinkle', 'grey', 'stubborn' and 'bingo' would make people walk slower. The person sitting in the corridor had measured how fast she covered the 9.75 metres from the lab to a 'broad strip of silver carpet tape' on the floor of the corridor.[30] The experimenter gives her a printed version of this explanation and thanks her again.

On average, people in the experimental condition did walk almost a second slower than those in the control condition (without the elderly words). None of the participants had been aware of anything unusual about the words. Their behaviour, in this case a slower walking pace, had been triggered automatically, bypassing consciousness. Bargh presented this result as a crucial advance in the study of 'the automaticity of social psychological phenomena'.[31] It was already known that, for instance, attitudes and stereotypes can be activated automatically by aspects of the situation one is in – a process known as priming – and that this effect may linger to influence the way new situations are perceived and evaluated. Although this in turn can influence behaviour, it was thought that behaviour was ultimately under conscious control. Bargh, however, hypothesised that behaviour too could be activated automatically by priming the mental representations of 'behavioral responses'.[32]

Among the historical precursors of this hypothesis, Bargh gave pride of place to William James's principle of ideomotor action: the representation of a movement awakens to some degree that movement itself. Via the work of Walter Dill Scott this principle had been a formative influence on young Dale Carnagey, convincing him of the power of suggestion.[33] To Bargh, the principle of ideomotor action was an important reason to think that priming a stereotype like 'the elderly' would have an effect not only on perception but also on behaviour because the stereotype comprises the representation of the behaviour of the elderly. Hence, reading words like 'forgetful', 'rigid' and 'gullible' could be expected to make people walk slightly slower. This field of research and its phenomena are often called 'social' or 'behavioural' priming to distinguish them from other kinds, such as semantic priming.[34] The latter is concerned with the effect of primes on the recall or recognition of associated words and is unaffected by the controversy surrounding social priming.

Bargh puts his work in the context of dual-process models: a family of psychological theories that describe behaviour as jointly produced by conscious control and unconscious, automatic processes. Priming experiments such as this one show that the latter are vastly more important than the former.[35] In Bargh's view, this is more than a matter of psychological theory. The extent to which thought and behaviour are under conscious control is also 'a fundamental existential question'.[36] The point of automaticity theory is not just that behaviour is mostly the result of automatic processes, many of which are triggered by a person's environment. It is also that this goes against the everyday misconception that we are in control of our own actions. Free will is an illusion.[37] Automaticity theory is a corrective; it has a moral-pedagogical message addressed to 'us': we are not as powerful as we think we are. But at the same it also shows that the psychologist *is* in control of our actions. At an aggregate level at least – experimental group versus control group – the psychologist can control how fast we walk, how aggressive we are and how we evaluate a social situation. What 'we' lose in rationality and agency, the psychologist seems to gain. Thus, priming research points to the possibility of a behavioural technology.[38]

With their claims of the power of unconscious influence, priming researchers find themselves in the vicinity of the territory covered in the previous chapter. For example, Johan Karremans, Wolfgang Stroebe and Jasper Claus found that priming people with a brand name (Lipton Ice, in their experiment) can affect their choice for that brand of beverage.[39] This may seem to indicate that subliminal advertising may finally become reality. The original claim, by James Vicary, that he could make people drink Coca-Cola by flashing the words 'Drink Coca-Cola' on the cinema screen, invited comparisons with brainwashing. Vicary never actually did the experiment – the story later turned out to be a hoax to attract attention to his marketing company – but subliminal messaging has continued to fascinate and scare people.[40] Wolfgang Stroebe was therefore careful to reassure his readers that his study with Karremans and Claus did not show that consumers can be brainwashed into buying Lipton Ice. Priming consumers in this way only works if they are already thirsty, and the effect is short-lived. But it does work, he insisted: we are constantly surrounded by a multitude of influences that may sway us this way or that, and psychology has discovered how to put these processes to good use. Priming is far less potent than what Vicary claimed for his subliminal advertising, but it is real science.[41] When Stroebe wrote this article, however, confidence in the robustness of priming effects had just been dealt a blow by the failed replication of Bargh's classic, genre-defining 'elderly walking' experiment.

Out of Control

In their replication study, Stéphane Doyen and his colleagues of the Free University of Brussels had decided to do away with the silver carpet tape and use infrared sensors, spaced 9.75 metres apart, to measure walking speed. This seemed to them more precise than manual measurement. In all other respects they were aiming for a replication that was 'exact', as they put it.[42] Nevertheless, they found no difference in walking speed between the experimental and the control groups. Searching for an explanation of Bargh's results, they hypothesised that they could be due to an experimenter effect: perhaps the experimenters in the original study had been aware of who was in the experimental condition and had unconsciously changed their behaviour towards those participants (with a slower pace of interaction, for example), and then the participants had unwittingly mimicked this by walking through the hallway more slowly – in other words, a priming effect but not the one that had been claimed by Bargh. Instead, the participants had been primed by the experimenters, who had been primed by the experimental material. Perhaps no one had been in control.

To test their hypothesis, Doyen and colleagues did a second experiment in which they shaped the expectations of the experimenters (randomly recruited students) in a 'one hour briefing and persuasion session'.[43] They also gave them a stopwatch to measure the participants' walking speed, telling them the infrared sensors were unreliable. The results partially confirmed their hypothesis: participants in the condition with the stereotypical elderly words now did indeed walk slower if the experimenter expected them to. The effect was the same whether speed was measured electronically or manually. But when the experimenter expected the subjects to walk faster, they didn't. The change in walking speed in the elderly condition was due to the combined effects of experimenter expectations and the elderly prime. 'Experimenters' expectations seem to provide a favorable context to the behavioral expression of the prime.'[44] Doyen and colleagues noted that this was in line with recent evidence, that showed 'that primed behavior is sensitive to the context in which it takes place'.[45]

Bargh was scathing about the replication study, but his arguments only underlined the sensitivity of priming effects. Doyen's experimenters, he noted, had directed the participants to the end of the corridor to make sure they crossed the two beams of the sensors and didn't wander. But this would have interfered with the priming effect: 'It has long been known, and shown in many studies, that calling the participant's conscious attention to an automatically operating process can reduce or eliminate that automatic effect.'[46] Secondly, Doyen appeared to have included elderly words in all thirty sets of words in the task, and this again would have alerted the participants to the prime and neutralised its effect. '[U]sing too many primes of the same category or theme can cause the participant to consciously notice that theme and, for the same reasons as above, eliminate or even reverse the natural priming effect.'[47] The automaticity of behaviour, which Bargh liked to emphasise so much as a corrective to our belief in free will and self-control, appeared to be easily upset by the slightest smidgeon of conscious attention.

Because of this sensitivity, priming experiments require precise control and great skill from the experimenters. This is the main message of the detailed description of priming methods that Bargh wrote with his student Tanya Chartrand.[48] Much of their handbook chapter deals with the prevention and detection of the subjects' awareness of the primes and/or their effect. The power of consciousness may be puny in comparison with the might of automaticity, but it is a nuisance all the same. Bargh and Chartrand describe how it can be dealt with by flashing the priming stimuli very briefly on the computer screen, too brief to notice – so-called subliminal priming. But one can also use supraliminal priming as long as the subjects remain unaware of the relation between the prime and the experimental task. In both cases, one is aiming for 'the "right" level of subtlety': the strongest possible manipulation, without making the subject aware of what is going on.[49] It is therefore essential to do an 'awareness check'.[50] In supraliminal priming one should use a so-called 'funneled debriefing procedure': a series of increasingly detailed questions after the experiment to find out to what extent the subject had been wise to the manipulation.[51] In subliminal priming one can simply present a few of the primes again and let the subject guess what they are. Whichever technique one uses, any subject who shows awareness must be excluded from the analysis.

Expertise

As the controversy wore on, reinforced now and then by new failures to replicate well-known studies, the fragility of priming effects and other social psychological phenomena only seemed to grow.[52] Commenting on the replication study by Doyen and colleagues, social psychologist Daniel Kahneman told Ed Yong that '[t]rivial details such as the day of the week or the colour of a room could affect the results'.[53] Ap Dijksterhuis thought this sensitivity was simply 'the reality of our subject ... many [social] psychological phenomena are affected by people's mood, by atmospherics, time of day, fatigue, motivation, and even the weather'.[54] As a consequence, if an experiment fails, it could be for any number of reasons. The simplest explanation is always that the replicator was incompetent. The Harvard psychologist Jason Mitchell set off a minor Twitter and Facebook storm when he formulated this standpoint in a paper that he published on his website.[55] '[N]ull findings cannot distinguish between whether an effect does not exist or an experiment was poorly executed, and therefore have no meaningful evidentiary value.'[56] Direct replication is useless: if the original effect is reproduced, we learn nothing new, and if it isn't, we learn nothing at all.

The expertise of the experimenter became a popular argument against nonreplications. Reacting to a failed replication of one of her studies, Simone Schnall, social psychologist at Cambridge University, presented herself as the victim of hostile and incompetent researchers. 'As anybody who's ever carried out a psychology study will know, there are many reasons why a study can go wrong,' she wrote.[57] Unfortunately, psychologists from different fields, who do not have the required expertise, think they can simply replicate your study by following the description in the methods section of the original article and then pretend that their inevitable failure falsifies the results that you have worked so hard for.[58] Neuropsychologist Matthew Lieberman concurred. Both Schnall and Bargh, he claimed, were the victims of unfair attacks by incompetent researchers. If replicators are biased against the original result and expect the experiment to fail, he argued, this will, among other things, influence how they interact with their subjects, which in turn will negatively influence the results. (Another experimenter effect, in other words.) We should be 'screening out' such people from replication efforts and only allow replicators who have shown that they can 'successfully get other priming effects, or other embodied cognition effects'.[59] Earlier, Kahneman too had contended that the subtlety of priming effects requires special abilities of the experimenter. Bargh, he said, 'has a knack that not all of us have'.[60]

The expertise argument didn't go down well with the proponents of replication studies. The anonymous blogger Neuroskeptic dubbed it the 'Harry Potter Theory of social psychology': apparently, it requires special skills ('magic') to produce such subtle effects, and if you cannot replicate them, it only shows you don't have these skills. You're a muggle, not a wizard or a witch.[61] Critics like Neuroskeptic argued that this makes scientific claims unfalsifiable: any nonreplication can be explained away by referring to unspecified skills that are needed to make the experiment 'work'. That's not how science works, added psychologist Andrew Wilson. All the relevant information required to replicate an experiment should simply be in the 'methods' section of the published report. Once you've made your work public, it's fair game for replication. There should be no requirement on replicators to be trained or vetted by the original researchers, as Kahneman, Schnall and Lieberman had suggested.[62]

Theory

Bargh had already realised that the knack was ultimately not a viable argument. Although he repeated that priming experiments require skill to make them work, he also emphasised that he 'never wanted there to be some secret knowledge about how to make these effects happen. We've always tried to give that knowledge away but maybe we should specify more details about how to do these things.'[63] He later conceded that the question why some people are able to produce priming effects and others are not is important in its own right because 'it draws needed attention to the precise contexts and conditions required to produce thoughts and behaviors from unconscious priming cues'.[64] These contexts and conditions had already become a concern before the crisis broke out. Since the 1996 elderly walking experiment, in which a prime appeared to have a direct effect on behaviour, a large number of moderators had been uncovered in priming studies. For example, a replication study of Bargh's experiment ten years later did find that priming the elderly stereotype made subjects walk slower, but only if they had a positive attitude towards the elderly. Those with a negative attitude walked faster.[65] Such variables had to be accounted for in a theory of priming. Bargh and his colleagues took the increasing number of nonreplications as an indication that there were more, as yet unknown moderators to be discovered. With their incompetent experiments the replicators had brought to light the need for further clarification of how and under what conditions priming works.

Ap Dijksterhuis maintained that 'the evidence for behavioral priming is overwhelming', since it has been demonstrated in hundreds of conceptual replications, but he did agree that the failed direct replications were 'a welcome, though rather loud, wake-up call'.[66] He dismissed the idea that they were proof that the original results were produced with questionable research practices. Instead, 'nonreplications of findings that have been replicated in many labs are much more likely caused by yet-to-be-identified moderators and boundary conditions'.[67] They showed the need for more precise theory so that the exact conditions under which priming effects can occur can be specified. This, he added, was particularly important in an applied context. 'Currently, behavioral priming is applied to increase fund raising, to reduce biases in elections, to improve eating behavior, and to help people to quit smoking and fight other addictions, to name just a few examples. Such laudable initiatives should not be jeopardised by too much skepticism, but there is no denying that a better understanding of the boundary conditions of behavioral priming will make applications more efficient.'[68]

Wolfgang Stroebe and Fritz Strack took a similar line: we should not be too worried by the much-publicised failures of direct replications of priming effects. Nonreplications can be due to 'an infinite number of influences that may come from the cultural circumstances, the experimental setting, characteristics of the participants, and so forth'.[69] Perhaps the participants in Doyen's experiment did not share the same stereotype of elderly people as the subjects in Bargh's original experiment. Stereotypes are culturally and historically variable. Simply copying the procedure of the original experiment does not guarantee the same result because the same stimulus may mean different things in different historical and cultural contexts. That is why social psychologists rely on conceptual replications, not on so-called direct replications. It is even possible, they speculated, that the fact that 'these findings are reported in most social psychology textbooks and are therefore widely known among student participants could have affected the results'.[70] Nonreplications are therefore not very informative in themselves: they should lead to a search for the 'moderator variables that determine the conditions under which theories apply'.[71] We do not need more replications; we need a better theory.

Joseph Cesario saw the fragility of priming effects in an evolutionary light. The variability and context sensitivity of priming effects is consistent with what we know of 'the evolved workings of the mind'.[72] In social situations, the brain not only reacts to the target stimulus but also takes into account 'information about the self, target others, and the broader contexts in which interactions take place'.[73] Thus, priming effects may be moderated by cultural differences and by aspects of the physical environment; they may depend on whether the individual is surrounded by other people, whether the subject is a high or a low 'self-monitor' and so on. Since priming research is still a young field, 'we do not yet have enough well-developed formal theories of priming that can pinpoint exactly what features and what effects should be important'.[74] As yet, any subtle, seemingly irrelevant change in the experiment can eliminate the priming effect.

All of which led Cesario to pose an intriguing question: if priming is so fragile, how had all those successful studies come about? Most likely as a result of good fortune, he thought. The impressive catalogue of priming effects was probably the result of researchers accidentally 'hitting on the right level of an important moderating variable'.[75] When they weren't so lucky and the experiment did not work, they simply discarded the results. '[T]his might be what is meant when researchers talk about having "insight" or "intuition" in conducting priming experiments in which they cannot verbalize why they made a decision but knew to make it. In the end, there is really nothing magical about this.'[76] However, Cesario urged his colleagues not to use the intuition argument to explain away nonreplications and to acknowledge the possibility that they point to problems with the original study, such as questionable research practices. Direct replications are important, and researchers should always try to replicate their own effects before publishing their work. 'Given the almost certain routine problems of our current practice (such as not being forthright about failed replications), conceptual replications without direct replications will only serve to perpetuate problems in the field.'[77]

Efforts were already under way to bring priming under control by describing its complexity in more refined theories. With a variety of moderators already to take into account, the task wasn't easy. Tobias Schröder and Paul Thagard presented a theory that integrates cultural, psychological and biological levels of explanation.[78] The mind, they posited, amalgamates all the representations that are active in it at a given time – including, for example, a primed concept and knowledge of the self, the surroundings and other people present – into a single 'Gestalt'.[79] This amalgamation is constrained by the culturally determined affective meanings of the represented concepts. The mind strives to satisfy these constraints in a consistent, coherent Gestalt of representations. Priming can then be conceptualised as 'a kind of parallel constraint satisfaction, with primes providing new inputs that produce outputs as the result of constraint satisfaction'.[80] The biological mechanism underlying it is provided by so-called semantic pointers, firing patterns in populations of neurons, a mechanism that also provides an explanation for the automatic translation of a Gestalt into behaviour. Although the theory is complex, it is still explicitly mechanistic. Schröder and Thagard finished by describing two computational models of their theory and simulations of priming that they performed with them.

With the nonreplications piling up, the theoretical effort gained urgency. Roberta Klatzky and David Cresswell's intersensory interaction model of priming was explicitly meant to 'shed light on issues surrounding the reliability of priming effects' that had arisen in the previous two years.[81] The goal, in other words, was to explain not just how priming works but also why it fails so often in the laboratory. Their model is an extension of a theory that describes perception as arising from the interaction of the sensory channels. Each channel (auditory, visual, kinaesthetic, etc.) puts in a 'bid' for the value of a particular property of the world; all the bids are weighted and then integrated into a perception. Other channels than the senses are also involved in this process. Information from memory may colour a perception by mediating the bids from sensory channels, for example. There is also heuristic bidding, where the value of one channel is inferred from the information from another channel or even the anticipated information from that channel. This same bidding model can be applied to describe priming effects: priming with elderly-related words activates the elderly concept, which puts in a bid for the 'self-judged energy resources', which in turn lead to slower walking.[82]

All is not well, however. Somewhat ominously, Klatzky and Cresswell write that '[a]s the intersensory integration model is extended beyond perception per se to memory mediation, anticipatory heuristic processing, and to nonphysical properties of the world, something is lost in the process, namely, determinism'.[83] Several sources of variability, including semantic context and conscious control, may influence these bidding processes to various extents, and all the channels are affected by noise and stochastic variation. The model suggests that priming effects are real but also difficult to replicate because they 'stem from inherently stochastic and context-dependent bases'.[84] Fortunately it also explains how the sometimes surprisingly large effect sizes in priming studies can come about. '[A]lthough the thrust of the present argument is to explain why social priming effects are vulnerable, the same approach also suggests why a range of effect sizes might be possible.'[85] They finish by offering some 'admittedly speculative ideas' on how their model could be used to study the moderators of priming effects.[86]

The more refined the model, the harder the task of making priming effects replicable seemed to be. Commenting on Klatzky and Cresswell's effort, Edwin Locke noted how complicated the models of priming had become and how many uncertainties were still left. 'Would a full theory of behavioral priming have to include thousands of moderators?', he asked rhetorically.[87] He did think it could be done, but it would require arduous, systematic theory building, for which he recommended a strategy of conceptual replications. Barry Schwartz, however, urged priming researchers to bite the bullet and give up the hope for a science of priming. What Klatzky and Cresswell had shown, he thought, was that in priming, anything is potentially relevant. 'People's associative chains are different, their cultures are different, what they happen to be paying attention to changes from one person to another and one moment to another. Almost anything can happen – even if priming is real.'[88] Priming effects may well be inherently unpredictable and the difficulties replicating them 'a feature, not a bug'.[89] Instead of stubbornly continuing to aim for controlled and replicable effects, we should try to understand the phenomena through 'a kind of history of social priming'.[90]

Replication and Technology

In the early days of social priming research, it appeared to promise techniques for moving people unawares, avoiding their possible resistance. With its large effects from subtle manipulations, moreover, it seemed to have a surprising amount of leverage. Priming appeared to be a powerful technology, a novel realisation of the dream of control. Of course, bypassing consciousness and achieving automaticity required careful calibration of the prime: it had to have enough force but stay outside of awareness. As powerless as consciousness was, it was imperative not to draw its attention to the prime. But if you got it just right, people could be primed in many ways; you could make them more creative by priming them with the Apple logo and more honest by priming them with Disney, for example.[91]

As conceptual replications of the basic priming hypothesis multiplied, however, the impression of efficient effectiveness started to fade. Researchers found themselves in a morass of moderators of their own creation. With each study finding an effect on behaviour of prime A but only in the presence of variable B, priming lost some of its crispness. Next, when more and more outsiders failed to replicate well-known results, the picture got murkier still. What had appeared to be a technique was now presented by priming researchers as to some extent an art that was still in the process of being formulated in terms of explicit procedures. Experiments in this field were said to require special expertise, intuition, a knack. Priming began to look like the perfect illustration of Quintilian's uncertainty principle, mentioned in Chapter 2: there is always an element of indeterminacy about persuasion and influence. The transparency and predictability that critics demanded were not possible yet. Moreover, the insistence that nonreplications were probably due to yet-to-be-identified moderators further compromised the technological promise of priming. If unknown moderators regularly spoil priming in the laboratory, then surely they will do so all the more in the frenzy of everyday life. Gradually, a world was unveiled that was not unlike the one described by Joost Meerloo fifty years earlier: a plenitude of interacting influences where no one is in control.

There is a competing account that says that perhaps this world does not exist at all. According to this account, when a direct replication of a social priming experiment fails, it's because there was no priming effect in the first place. The original result was the product of questionable research practices. For example, personality psychologist Sanjay Srivastava has argued that there is compelling evidence that makes the unknown moderator explanation of nonreplications dubious.[92] In the so-called 'Many Labs' replication projects, a number of classic and more recent studies were selected for direct replication, and the replications were subsequently performed independently in a number of laboratories.[93] If the fabled unknown moderators are an important factor in replication success, one would expect results to vary between laboratories, but that was not the case. An original result could either be replicated by every lab or by none of them. The variability predicted by the unknown moderator account wasn't there.

That appears to leave questionable research practices as the only other explanation for the results of studies that could not be replicated.[94] But it needn't be intentional malpractice; instead the results could be due to data-dependent analysis, what Andrew Gelman and Eric Loken have called 'the garden of forking paths'.[95] A researcher need not explicitly fish for statistically significant results by trying out many statistical tests on multiple variables until one 'works', but if her choice of the one test she performs is taken after looking at the data, whatever statistically significant result she finds is still likely to be noise. If the data seem to indicate that variable A has an effect on variable B but only in the presence of variable C, and that makes sense theoretically, it may seem natural to test that, and only that, relation. But other paths through the data could have been taken that might have been equally plausible in light of the theory.

Statistically speaking, data-dependent analysis is particularly risky, Gelman and Loken write, in a research context of 'small effect sizes, small sample sizes, large measurement errors, and high variation', which tends to be the case in social psychology.[96] But those are also the kind of circumstances in which researchers are prone to data dependent analysis. 'Nobody wants to starve. If there's no fruit on the trees, people will forage through the weeds looking for vegetables. If there's nothing there, they'll start to eat dirt. The low quality of research in these subfields of social psychology is a direct consequence of there being nothing there to study. Or, to be precise, it's a direct consequence of effects being small and highly variable across people and situations.'[97]

There are other positions regarding the reality of priming effects, but whichever account one chooses, it seems unlikely that priming currently offers an efficient and effective behavioural technology. Even if priming effects are real, it is difficult to see how any reliable, real-world applications would be possible if the effects can be moderated by many factors, including an unknown number of yet-to-be-identified individual and contextual variables. The severity of the replication crisis, particularly in priming studies, puts in doubt the technological aspirations of this field of research. The philosopher Hans Radder, developing ideas of Jürgen Habermas about the relation between science and technology, has argued that a science shows its technological potential in the reproducibility of the material realisation of its experimental processes, what psychologists call direct replication. 'Just like experiments, working technologies need to be stable and reproducible, while the control of the relevant interactions between the technological system and its environment constitutes a necessary condition for achieving this goal.'[98] A stable and reproducible experiment can be a first step on the way to a working technology, although Radder notes that technology is typically required to remain stable in a larger variety of contexts and over a longer period of time. '[W]e cannot assume that a successfully realized experiment guarantees the success of the corresponding experimental technology.'[99] If there isn't a reproducible experiment in the first place, however, whatever putative technology is out there in the real world is not backed by any achievements in the laboratory.

Trials of Strength

Despite its apparently uninteresting nature of mere repetition, over the past few years direct replication has proven to be a productive, eventful research strategy in psychology. Direct replications have made things happen in the discipline: they have instigated theoretical developments and have fuelled debates about fundamental issues in research methodology and statistics, publication practices and communication between researchers. The various Open Science initiatives, arguably the most important infrastructural changes in the discipline of recent times, not only facilitate replication studies but in turn get their urgency and visibility in part from the concerns raised by failed direct replications. Direct replications have spurred a wave of innovation in psychology.

There remain detractors, who insist that the inherent complexity and context sensitivity of psychological phenomena make direct replication generally unfeasible. Instead, conceptual replications are the right research strategy, especially in basic research, where the concern is not with the reliability of a particular experimental procedure but with the underlying theory. Apart from the fact that this leaves the possibility of effective real-world applications in doubt, it seems questionable to me whether the reliance on conceptual replication has been such a fruitful strategy for theoretical development.[100] Conceptual replication has an inherent confirmation bias because unsuccessful replication studies tend to be dismissed as apparently imperfect operationalisations of the theory's constructs. Thus, in practice the theoretical import of conceptual replications lies mainly in the addition of moderator variables, as the priming saga illustrates so well. The theory is never put to the test; it only becomes more and more complex. Michael Billig pointed out long ago that this process tends to repeat itself in social psychology: once a theory has been fragmented into a multitude of problems of restricted range, researchers lose interest and shift their attention to a fresh, new topic, and the process of complexifying it starts anew. 'The net result is that the more comprehensive textbooks become ever more complex and their writers claim that this complexity is the real discovery of social psychology.'[101]

A more radical argument is that the many failed replications show that the use of the experimental method as such is problematic in psychology. Radder, for example, puts the blame for the current problems in psychology on the discipline's reliance on experimentation.[102] The replication crisis is not 'a problem *within* experimental methodology, but . . . a problem *of* the experimental approach as such'.[103] According to Radder, the so-called 'double hermeneutic' makes it doubtful that experimentation can lead to generalisable knowledge in psychology.[104] The term 'double hermeneutic', introduced by sociologist Anthony Giddens, refers to 'the fact that the outcomes of experimental research on human beings depend not only on the interpretations of the scientists but also on those of the research subjects'.[105] Radder argues that this gives the socio-cultural context of an experiment an epistemic significance that 'reveals the limitations of an individualizing experimental psychology, in which the social is no more than the statistical aggregate of the behavior of individuals'.[106] Similar arguments have long been put forward by critics of experimental psychology. For example, in 1973, during the previous crisis of confidence in social psychology, Ken Gergen ignited a major controversy when he argued that social psychology cannot be a predictive science, nor should it strive for a technological ideal.[107] Human social behaviour is culturally and historically variable, and psychology is itself one of the contextual variables, to which people may react in order to resist what is predicted of them. As a result, social psychological phenomena are 'largely nonrepeatable'.[108] Social psychology, Gergen concluded, should accept the fact that it is a form of history.

When Barry Schwartz, quoted above, proposed 'a kind of history of social priming', he referred to Gergen's work, but the impact that such arguments have had on the way experimental social psychology is practiced has so far been rather limited. The contextual and even the historical variability of behaviour is widely acknowledged and is said to make it a uniquely difficult topic of investigation and psychology 'the hardest science'.[109] But underneath that variability are thought to be generalities, the properties of a context-sensitive mechanism that can be studied experimentally. As Barry Schlenker argued in his reply to Gergen's 'Social Psychology as History', all scientific propositions are conditional, but if they are sufficiently abstract there is no reason why they couldn't be generalisable.[110] Psychologists therefore generally do not see contextual variability as a fundamental problem. As long as one relies predominantly on conceptual replication, that variability simply offers a rich terrain of experimental study along the lines sketched above: starting with a simple, straightforward theory and proceeding to explore its contextual complexity.[111]

Like Radder, I believe that the replication crisis shows the limitations of this strategy. In fact, it seems to me that the turn to direct replication has done more to upset this practice than decades of philosophical arguments. Suddenly, the fragility of phenomena has become a nuisance. Whereas the strategy of conceptual replication allowed researchers to discover moderators at their leisure, now every failed direct replication comes with the implicit or explicit suggestion that there is another, as-yet-undiscovered moderator at play, the experimental procedure needs further refining or the original result was the product of sloppy research. In any case there is work to do for the original researchers and their field – that at least is the conclusion often drawn. Of course, a spate of nonreplications is not the conclusive, incontrovertible argument that experiments just do not work in a particular field of study, just like a single nonreplication does not mean the original effect does not exist.[112] The point is that the much-vaunted complexity and variability of behaviour are no longer a source of results at the disposal of the researcher but have become an active force in the research process. Paradoxically, the demand that a researcher must control a phenomenon to such an extent that she can reliably reproduce it according to a particular procedure has given an opportunity to whatever resists that control to make itself apparent.

In Chapter 2, I discussed Latour's proposal that the scientific research process should be a trial of strength in which, to make it productive, the object must be allowed and encouraged to object to our propositions. He believes that the social sciences, including psychology, focus too much on controlling their subjects, and as a result the knowledge they produce mostly lacks the surprising novelty that the natural sciences are so good at.[113] I think the situation is more complicated than that. As critics like

Brian Nosek have pointed out, surprising novelty has become almost a sine qua non for publication in high-impact journals, and social psychology in particular excels at the production of counterintuitive, 'sexy' results. Their novelty value, however, often lies in the message that people are not as much in control of their lives as they think and not as rational and as moral as they pretend to be. Simultaneously, the power of the psychologist, who is able to manipulate the subjects into their automatic, irrational and immoral behaviour, is affirmed. What the events of the past five years seem to indicate is that that power depends to some extent on permitting oneself a certain amount of flexibility: avoiding the stringent demands of direct replication, in the first place, and, in some cases, exploiting researcher degrees of freedom and roaming the garden of forking paths. That flexibility is now being curtailed by the turn to direct replication, by pre-registration of hypotheses and data analysis plans and by data sharing. As a result of these constraints on the researcher, the power of the object of study to resist the propositions of the researcher increases in equal measure. The unpredictability and variability of human behaviour that have long been argued on theoretical grounds now have a chance to really trouble the psychologist.

At the same time, direct replication in turn tests the limits of human unpredictability and recalcitrance. Human behaviour is not uniformly sensitive to context, people's own interpretations of the situation do not always have an influence on how they react, and their behaviour often fails to conform to their intentions. Human behaviour is historically and socially variable but to varying extent. In other words, the variability of psychological phenomena is itself conditional, not absolute. As a consequence, the replication crisis does not affect all areas of psychological research equally, and within social psychology the vulnerability varies as well. As direct replication becomes more common, it will gradually become clearer which psychological phenomena lend themselves to an experimental approach and to what extent. A blanket rejection of the experiment does not give whatever is stable and generalisable about human behaviour a chance to assert itself.

10 Conclusion

The histories of human engineering in this book present a number of forms that the dream of control has taken with regard to human behaviour and society. I could have chosen many others: psychotechnics, for example, or social engineering, both of which I have briefly touched on. Their proponents were upfront about their ambition to emulate the engineering of the material world, as was B. F. Skinner, whose behaviourism is perhaps the clearest example of a technology of control in psychology. I could also have gone further back in time, to the work of the industrialist Robert Owen, who believed among other things that '[c]hildren are, without exception, passive and wonderfully contrived compounds; which, by an accurate previous and subsequent attention, founded on a correct knowledge of the subject, may be formed collectively to have any human character'.[1] With the histories in this book I do not pretend completeness in any way, even for the twentieth century. I have chosen them because they show a dynamic that I believe is typical of such calculated strategies of control: they have to contend with resistance from the people they deal with and with incalculable aspects of the situation, and they do so with what I have called 'tact', actions that accommodate the sensitivities of the people involved and exploit what is opportune in a situation. This has been the primary dynamic in this book, but it is connected, in various ways, to varying extents, with other oppositions that run through these cases.

First of all, these technologies come with an appeal to science and with an implicit or explicit distinction from unscientific and therefore ineffective alternatives. Albion Small demarcated his social technology from the benevolent amateurs of the Social Science Association. Social reform should be based on thorough knowledge of the laws that govern the social process, not on good intentions and surveys. The Social Science Association had itself been set up to make social reform scientific, and after Small and Henderson other sociologists would again redraw the boundary between rational, objective science and the work of do-gooders. In the studies of Russian and Chinese brainwashing in the

1950s, on the other hand, a distinction between scientists and evil amateurs played a role. An important question was whether the techniques of brainwashing had been developed by scientists (pupils of Pavlov, for example) or were just the result of centuries of trial and error by the executors of tyranny. In the latter case, there might be room for scientific improvement of mind control. At other times, science was contrasted with organised politics rather than with the efforts of amateurs. Again, Small's social technology is an example, as he presented it as an alternative to socialism. The later social engineering movement emphasised its superiority to politics even more, and in a way it was also the point of Frederick W. Taylor's scientific management. Negotiating with the unions was not a rational way of setting wages and determining working conditions. Taylorism offered a scientific alternative.

Secondly, the scientifically informed expertise of the human engineers is contrasted not only with unscientific alternatives but also with the ignorance of the people they are dealing with. Frederick Taylor thought workers were incapable of understanding the science behind their work, and he institutionalised that ignorance by separating the planning and execution of work. From then on, the managers had objective, scientific knowledge that the workers were not privy to. The workers simply had to follow the instructions of the manager, doing the job exactly as had been shown in the time that had been calculated for it. Other technologies exploit mechanisms that ordinary people cannot know about because they're hidden in their own minds. These strategies work so well precisely because the subject is clueless. The unconscious played an important role in several histories. In theories about brainwashing, such as those of Joost Meerloo and Robert Lifton, mind control was thought to work in part through the manipulation of unconscious fears and feelings of guilt. In Dale Carnegie's work as well as recent priming research, the technology targets unconscious, automatic links between ideas or emotions and behaviour, connections that William James had described in the law of ideomotor action. The claims of neuromarketing that I briefly described in Chapter 1 are built on the idea that neuroscientists, with the help of EEG and brain imaging technology, can finally see the real mechanisms that underlie behaviour. These mechanisms are inherently invisible to laypeople; their ignorance is a simple consequence of the fact that they cannot look into their own brain.

Strategy

These two oppositions, between science and non-science and between scientific knowledge and the ignorance of the people it is about, both express the idea that science describes a deeper, more important reality than either its competitors or its subjects can. That reality becomes the domain of strategies of control, based on procedures and principles derived from the theory. Taylor's domain was the factory floor, where individual workers, machines and tasks were combined into efficient production processes. Albion Small's social technology operated in a society that was described in terms of interests, associations and goals, whereas that of Karl Popper was composed of institutions. The social technology of Kurt Lewin and his students targeted groups, which were theorised as dynamic force fields. The world of Dale Carnegie, of Joost Meerloo, and of priming research is one of individuals and the influences that work on their unconscious.

In several cases, the strategies in this domain are literally calculated. Taylor's approach to management was a perfect example: he broke up work into elementary tasks and the gangs into individual workers, measured the time that a first-class man required for each task and calculated the one best way to do a job and how much the worker should be paid for it. In Kurt Lewin's work, the mathematics of the life space allowed, in theory at least, both predicting when change would be resisted and devising ways to overcome that resistance and move the individual and the group to a new equilibrium. Schröder and Thagard's theory of priming is a recent example of literal calculation: the mind is modelled as a space where representations, including primes, are amalgamated into a single 'Gestalt' in such a way that the constraints set by the culturally determined meanings of the representations are maximally satisfied. The process can be computationally simulated. In time, such computations might direct practical interventions.

In other cases, the outcome of the transformations that the theory allows is not numerical, but it is nonetheless derived from the theory. For Small and Henderson, sociology is the theoretical space where knowledge of all the sciences relating to people is integrated and transformed into social technology: rational plans for social reform. Dale Carnegie's numbered lists of principles for dealing with people followed straightforwardly from the self-interested, self-centred nature of human beings and the dominance of their unconscious. Edward Hunter described brainwashing as the 'calculated destruction of men's minds': it was a set of procedures that had been perversely derived from psychiatry and psychology, for example, by inverting the process of psychotherapy.[2] In Karl Popper's work, finally, science is itself technological, and engineering is the practical application of that technology. Knowledge of natural or social laws is the product of piecemeal, problem-driven tinkering in the laboratory, and that knowledge can then be applied by engineering machines or institutions.

Tact

Resistance introduces complications in the simple pairs of science and non-science, knowledge and ignorance. New oppositions appear, and incalculable factors enter the equation. The resistance to scientific management confronted its strategies with 'the human factor'. A contrast was constructed between the automatons that would be the inevitable result of Taylorism and the human beings that the workers should be allowed to be. The difference was 'the human factor', formulated with a range of concepts (initiative, soul, free will, individuality and so on) that expressed that people are not machines. Taylor defended his system by saying its mechanism is not its essence: scientific management can only work in an atmosphere of camaraderie and the acknowledgement of shared interests. Around the precisely calculated, strictly mechanistic work process there had to be a sphere of friendly human relations. Thus, in Taylor's view there remained a neat separation between calculation and the human factor: there was nothing human about organising the production process and nothing calculated about the friendly atmosphere on the shop floor. It was a square deal, but neither the unions nor the congressional committee trusted Taylor's offer. I've pointed to Mary Parker Follett as one of the first management thinkers to propose that the possibility of resistance had to be dealt with in the techniques of management themselves. Simply giving orders will always run up against the human need to be self-governing, no matter how friendly the boss is after work. Instead, the order had to be delicately placed 'in the situation' for both boss and worker to see and follow. Tact becomes part of management; the human factor is a strategic factor.

The idea of self-governance also complicated the relation between science and politics. Whereas the social engineering of the 1930s aimed to replace politics with a scientific, rational organisation of society, the disastrous consequences of communism and fascism prompted some to rethink the social role of science. Both Karl Popper and Kurt Lewin saw science as crucial in the functioning of democracy but in very different ways. In Popper's work, science and democracy are virtually indistinguishable: democracy must be scientific, a continuous process of piecemeal, hypothesis-driven social experimentation and critical evaluation of the results, and science for its part must be democratic in the sense of transparent and critical instead of dogmatic. But whereas the distinction between science and politics is thus partly erased (partly, because normative considerations fall outside the scope of science), Popper emphatically prohibited the social engineering of the personal element. Like Taylor, Popper neatly separated science and the human factor. Science was the sphere of rational decision making, but the personal element was an irrational, incalculable factor in society and should remain so. Engineering individual preferences and opinions, or moulding personal relations, would invalidate social experimentation because there would no longer be an independent criterion by which to measure its success. The human factor must remain outside our social technology, irrational and free, because otherwise we open the door to tyranny. There is no need for tact in Popper's scheme because there is, theoretically at least, no contact between technology and recalcitrant, incalculable individuals.

Kurt Lewin and his students made science serviceable to democracy in a different way. Lewin was as concerned with democracy as Popper was, but the techniques of democratic leadership that he and his students developed linked self-governance (democracy) and calculated, strategical guidance (leadership) into a paradoxical combination. These techniques worked by making the group the agent of change and by delicately guiding it towards the discovery of facts and solutions placed in its path by the group leader, not unlike Follett's demands of the situation. These were techniques that prevent resistance by erasing themselves. There is something deceptive about Alex Bavelas, Lewin's charmer: we see what he is doing, and Lewin can describe it in terms of field theory, but the group does not realise what is happening.

In the world of Dale Carnegie, in mind control and in priming research, the relation between science and the unconscious is complicated primarily by the consciousness of the subject. Human behaviour may be largely the product of automatic, unconscious processes but not entirely, and manipulating these automatisms works best if the target does not notice or is otherwise put at ease. Successful priming, it appears, depends on making the prime forceful enough to trigger a representation but not so forceful that it crosses the threshold of awareness. One should also make sure not to rouse the participant's suspicion in other ways; don't be too emphatic in your directions, for example. In priming research, it is essential to stay under the radar. In contrast, Dale Carnegie's techniques for dealing with people were not invisible – indeed, everyone could read about them in the book – but they were also, at the same time, non-technical. These techniques were manipulative, but they could only work if the resistance that manipulation provokes could be prevented by being sincere. The hybrid self this required, a self both instrumental and sincerely appreciative of the other, proved to be attractive to many, but it was revolting or ridiculous to those who wished to uphold a distinction between true individuality and 'dealing with people'. In the KUBARK manual, the fruit of a decade of mind control research, a Carnegie-like tact is recommended to create rapport with the interrogatee. Studies of brainwashing and other forms of mind control had consistently shown that insofar as such techniques worked, it was to a large extent because they involved a personal relationship between interrogator and victim. Interrogation, like brainwashing, proceeded within boundaries set by coercion, but the engine of the transformation that took place (adopting a new ideology, providing information) was the relationship that the interrogator tactfully created in that situation.

Enacting the Human

In my discussion of Latour in Chapter 2, I argued that, when analysing histories of human engineering, we should not ascribe to human beings certain characteristics in advance. It is more fruitful to consider these characteristics as the (provisional) outcome of ongoing trials of strength. Strategies of control and the resistance they provoke produce versions of being human. These versions are not just theoretical; they are more than descriptions. For example, the Taylorised workshop created a particular kind of worker, one who shouldn't think about the job but accomplish it with predetermined, precise and regular movements. In reaction, the unions mobilised another kind of worker, one who showed his initiative, free will and individuality by rejecting scientific management and going on strike. Lillian Gilbreth in turn tried to incorporate this worker back into the strategic domain. Her psychology of management focused on the individuality of the workers, which could be measured with psychological tests and nurtured by giving workers feedback on their output compared with that of others. It offered workers a way of being unique individuals within the context of a scientifically managed workplace. For all concerned, this was more than mere theory. It was, as Dale Carnegie would have put it, a new kind of life.

Tact is equally concrete and productive: being tactful is performing a version of being distinctly human and getting the other to play along. When Taylor rubbed shoulders with the workers on the shop floor, being perfectly straightforward with the men but kind at the same time, the camaraderie created right there and then was one version of being human, not a machine. It was a very masculine type of human. Taylor was one of the boys, getting his hands dirty, and all were united in the pursuit of more production, profit and income. Dale Carnegie's smiling, sincere appreciation enacted a different kind of human being: a friend who really cares.

The other is still made to be more than an automaton and mutual profit is still the goal, but this is a more polished human being than Taylor's. It is a white-collar version. This one remembers your wife's birthday. There are glimpses of this kind of humanity in studies of brainwashing and in the KUBARK manual. Finally, in Alex Bavelas's work, whether it was training leaders, changing eating habits or solving conflicts on the shop floor, a human being was enacted who was reasonable and objective and willing to work with others. She is a constructive group member who wants to solve the problem together rather than waiting for orders. This human being is a democrat, a responsible citizen, committed to the common good.

Principles and Examples

Tact prevents resistance by performing a relationship between two or more people that is not strategic. How to do so can to some extent be articulated in the form of techniques and principles – that is part of what makes it paradoxical. Taylor offered several techniques for bringing about the required mental revolution, including various communication strategies. Carnegie presented numbered lists: three fundamental techniques of dealing with people, six ways to make people like you, twelve ways to win people over, seven rules for a happy marriage. Cartwright spelled out eight principles of overcoming resistance to change, and the KUBARK manual gave a number of suggestions for creating rapport with a potentially recalcitrant interrogatee. Tact is not entirely ineffable; it can to some extent be explicated.

There remains a gap between theory and practice, however: knowing the principles of tact is not enough to be successfully tactful in actual situations. The principles act as guidelines that need to be adapted to the situation at hand and the people concerned. When teaching tact, the gap between theory and practice is bridged by examples. Machiavelli explained what a sense of 'occasione' consists in by showing how the great rulers and warriors of old had seized the opportunities that had presented themselves. Similarly, Taylor and Carnegie were consummate raconteurs with a repertoire of stories and anecdotes to illustrate what tact looks like. Tact is taught primarily by principle and example rather than by theory and script. The principles do not precisely determine how one should act in each particular situation; they rather supply general directions – what is the problem to be solved, where lies the solution, what to avoid and what to seek – and give a vocabulary to talk about problem, situation and solution. The examples translate the principles into concrete actions, but the principles do not provide algorithms for generating specific actions and scripts. The translation has to be made anew each time in every new situation. Part of being tactful is having a sense of *kairos*, of opportunity and how to seize it, and you acquire it not by memorising procedures but by learning and discussing principles, by observing examples and by practising.

Consider, for instance, how Bavelas taught democratic leadership. Noteworthy is that in his lessons, the example would precede discussion of the principle that it expressed. First of all, Bavelas used his own way of leading the group as an example of the kind of democratic leadership he wanted to impart. He would draw the group's attention to the way he had approached a particular situation and then have the group members discuss what they had seen. Other examples were furnished by films of the group members' own efforts or of the experiments on leadership styles of Lewin and Lippitt, and each time there followed a discussion to work out what principle that particular case was an example of. Above all it was important to teach the students that the techniques should never be used as a set of tricks but should always be adapted to the situation at hand. Thus, principles and examples were equally important: the examples gave the principles concrete meaning, but without the principles the examples had no point.

The recent replication debate revolves in part around this same issue of a gap between theory and practice. One argument against direct replication made by people like Stroebe and Strack is that one cannot expect the same experimental procedure to give the same result in every context. What the theory means, how its terms should be operationalised, depends on the historical, social and personal aspects of the situation. What worked in the USA may fail ten years later in Britain with a different sample of students. For each particular context, the theory has to be translated again into a specific procedure. Replication only makes sense as conceptual replication. Thus, theory and replication relate to each other as principle and example. Each successful experiment exemplifies what the theory means, but there is no single, unique procedure that leads from the theory to a predicted result. Translating the theory into an experiment that 'works', some say, requires 'a knack', 'intuition' or 'expertise': some form of tacit knowledge not (yet) articulated in the methods section of the original study. To the proponents of direct replication, this is unacceptable: it reinforces the confirmation bias in psychology, making theories virtually unfalsifiable. If the theory does not fully determine what should happen in an experiment, then it can never really fail. Besides, intuition has no place in science. Whatever is tacit needs to be made explicit. If the theory does not predict exactly what will happen in a particular situation, it is simply not robust and detailed enough, and it is up to the original researchers to come up with a better version and test it.

Science Fiction

The point I've tried to make in this book is not that there are no useful applications of social science or that social scientists are not doing important work in society. It is, rather, that their techniques are accompanied by tact: that these techniques require judgement of the situation in which they are put to work and that their technical character needs to be made palatable or invisible to prevent resistance. The importance of *kairos* complicates the opposition of science and ignorance that human engineering likes to play on. Tact is crucial to the functioning of these strategies, but it is not itself an entirely calculable strategy. Tact introduces an element of ignorance: you cannot fully know how to do it; you have to develop a feeling for it. It is to some extent a tacit form of expertise. This sits uneasily with the dream of control; it implies human engineering has not attained the Platonic ideal of a *technè* because it cannot spell out exactly which causes move people in which way in which circumstances.

I do not take a principled stance on whether a tactless *technè* will ever be possible.[3] I prefer to approach the issue empirically and study the continuing history of that question in attempts at human engineering, in which the extent to which human behaviour can be controlled in a predictable fashion without having to deal with the resistance of the people involved is being tested. If we look at the latest, cutting-edge attempts, we do not seem to be there yet. In the first chapter, I introduced two attempts to elevate human engineering to the level of a *technè* and make it predictable, effective and efficient. Priming at first appeared to be a technique that was effective because it remained under the radar of awareness, covertly targeting the automatisms in the mental apparatus. It now turns out to be a bit too fragile, even in the rarefied atmosphere of the laboratory, to offer a predictably efficient and effective behaviour technology. At least for the time being, that fragility makes it unlikely that priming techniques fulfil their promise of predictable effectiveness in the real world. Indeed, the frequent references to a need for special expertise, made in connection with priming research, can be read as acknowledgements that noncalculable judgement plays a large role in the production of social priming effects even in the laboratory.

What of that other technology mentioned in the introduction to this book, neuromarketing? Will the turn to neurotechnology finally make the human factor irrelevant? Will it free the human engineer from the need to pander to the human illusion of control in order to get the job done? Does targeting the brain directly make it possible to bypass the recalcitrant subject and do efficiently and effectively what is good for him or her or for the client? For now, there remain doubts about the extent to which the various techniques of neuromarketing fulfil their promise. In a 2011 white paper prepared by the Advertising Research Foundation, academic experts in psychology, marketing and cognitive neuroscience raised a number of concerns regarding the methods used in neuromarketing.[4] They noted among others that sample sizes tend to be small, studies are done without a control group, and there is not enough attention to the validation of constructs and measures. It doesn't help that most of the techniques are proprietary and that there is little transparency about the methods. The experts also thought that the field was being oversold. 'Exaggerated claims have promoted the belief that people are entirely driven by emotion, that neuromarketers have access to consumers' "subconscious" desires and that they can manipulate those desires against their will. We did not find support for such beliefs in this project.'[5]

But let us assume, for argument's sake, that neuromarketing researchers in due time do develop reliable, effective and efficient techniques that can precisely measure the consumers' subconscious reactions to advertising material, thus allowing advertisers to manipulate the consumers' desires, even against their will. A posthuman era has been ushered in, in which consciousness and rationality have become irrelevant. In marketing at least, what people think and say about their preferences is no longer of any consequence whatsoever. The neuromarketer has become an engineer who needs to possess no people skills beyond basic good manners. Tact has disappeared from human engineering. In such a world, I believe, people would need convincing that it is to their advantage to consent to being manipulated, to cooperate and not to resist human engineering as such. I think neuromarketers realise this. They know that when words are no longer necessary when inspecting the consumer's subconscious, they are all the more needed to persuade citizens that such a practice is to their benefit as well. It is not intuitively attractive. That is why they write books such as *The Buy Button* and *Free Will Doesn't Exist*. Even tactless forms of human engineering need to be wrapped in skilful rhetoric.

Notes

1 Introduction

1. De Munnik, *De koopknop*.
2. Ibid., 25.
3. Mönning, 'Neuromarketing: De "koopknop" is een illusie'.
4. Historian Stefan Schwarzkopf has shown that this revolution is itself not new: the promise of more effective marketing by targeting the brain goes back to the late 1960s. Schwarzkopf, 'Measurement Devices and the Psychophysiology of Consumer Behaviour'.
5. Lamme, *De vrije wil bestaat niet*.
6. Dijksterhuis and van Knippenberg, 'The Relation between Perception and Behavior, or How to Win a Game of Trivial Pursuit', 868–69. In another condition, soccer hooligans replaced secretaries.
7. 'D&B – wetenschappelijk inzicht & gedragsbeïnvloeding in de praktijk'.
8. Schneider and Woolgar, 'Technologies of Ironic Revelation'.
9. As Schneider and Woolgar put it: 'The revelation is ironic in the technical sense that justification for the use of these technologies depends on a constructed incongruity between what is expected and what actually is the case.' Schneider and Woolgar, 'Technologies of Ironic Revelation', 171.
10. For historical studies of such boundary work, see, for example, Coon, 'Testing the Limits of Sense and Science' on Joseph Jastrow and my own study of Johannes Linschoten, Derksen, 'Are We Not Experimenting Then?'
11. Neurensics and dijksterhuis&vanbaaren, 'Front Runners in Neuro-Economie Kondigen Alliantie Aan'.
12. Cited in Vickers, *In Defence of Rhetoric*, 203.
13. Van Baaren in 'Een luisterend oor is het halve werk', 6. The idea that people are naturally resistant to change was criticised in the 1950s by Douglas McGregor 'The Human Side of Enterprise', who saw it as a core tenet of the control-oriented approach to management that he called 'Theory X'. Dent and Galloway Golberg, in a recent critique, note that the idea of resistance to change is still 'almost universally accepted in organizational life'. 'Challenging "Resistance to Change"', 25.
14. Van Baaren in 'Een luisterend oor is het halve werk', 7.
15. 'Het Nederlandse standaardwerk voor overtuigen'. Van Asperen de Boer, 'Herziene editie ijs verkopen aan Eskimo's'.
16. 'Zij ontdekte de wetmatigheden van het overtuigen'. 'Lezingen en workshops'.

17. Van Goethem, *IJs verkopen aan eskimo's*, 24.
18. Ibid., 35.
19. Ibid., 60.
20. Ibid., 61.
21. Ibid.
22. Somewhat surprisingly, Van Goethem includes Margaret Thatcher in her list of maximum persuaders: she was an authority during her reign and additionally became an example in later years and even a friend after the movie *The Iron Lady* with Meryl Streep. The decidedly mixed reactions to Thatcher's death speak against Van Goethem's interpretation.
23. Ibid., 76.
24. Ibid.
25. De Certeau, *The Practice of Everyday Life*, 37.
26. Or not yet their own: the sometimes slapdash efforts of first-year psychology students when they participate in experiments in exchange for course credit are a tactic that is analogous to la perruque.
27. Ibid., 24.
28. Ibid., xxiii.
29. Bricolage or 'tinkering' in science and the tactics of the users of technology have been central to the study of science and technology over the past three decades.
30. Ibid., 73.
31. Philosopher of technology Andrew Feenberg has used De Certeau's concepts of strategy and tactics to sketch a way 'beyond both dystopian condemnation and the posthumanist acceptance of technology', Feenberg, 'Looking Backward, Looking Forward. Reflections on the Twentieth Century', 104. The way technology is adopted, adapted and resisted in everyday life shows that technology is not the totally oppressive, controlling force that dystopians see in it nor as seamlessly integrated in our being as the posthumanists think.

2 Tact and Technology

1. An encomium is a speech of praise. Gorgias, however, constructs an apology of Helen's actions rather than praising them.
2. Gorgias, *Encomium of Helen*, 29.
3. Ibid.
4. Ibid., 27.
5. Ibid.
6. Ibid., 31.
7. Ibid., 23–24.
8. McComiskey, *Gorgias and the New Sophistic Rhetoric*, 28.
9. By Phillip Sipiora, in his introductory chapter in Sipiora and Baumlin, *Rhetoric and Kairos Essays in History, Theory, and Praxis*. The recent revival of interest in *kairos* has produced an extensive literature. I have drawn also on Paul, 'The Use of Kairos in Renaissance Political Philosophy'.

10. See Carolyn Miller's foreword in Sipiora and Baumlin, *Rhetoric and Kairos Essays in History, Theory, and Praxis.*
11. Cicero, *Brutus, Orator; Volume V*, 359.
12. Plato, cited in Billig, *Arguing and Thinking*, 84.
13. Ibid., 92.
14. Ibid., 62.
15. Ibid.
16. Miller, 'Should We Name the Tools?'
17. *Institutio Oratoria*, 11, 3.
18. Kennedy, *The Art of Persuasion in Greece.*
19. Plato, *The Collected Dialogues of Plato*, 4.
20. Ibid. Socrates's defence was reported by Plato in his 'Apologia'. Kennedy considers it doubtful that Socrates ever spoke these words. Kennedy, *The Art of Persuasion in Greece*, 149.
21. Sloane, *Encyclopedia of Rhetoric.*
22. Miller, 'Should We Name the Tools?,' 20. Foucault noted that the same is true of power: 'power is tolerable only on condition that it mask a substantial part of itself. Its success is proportional to its ability to hide its own mechanisms'. Foucault, *The History of Sexuality. Volume 1: An Introduction*, 1:86.
23. Miller, 'Should We Name the Tools?,' 28.
24. Quoted in ibid.
25. 'Onzichtbaar Overtuigen'.
26. Hogan and Speakman, *Covert Persuasion: Psychological Tactics and Tricks to Win the Game.*
27. 'Vertel aan geen enkele vrouw dat je met *MasterFlirt* bezig bent. Je zult zien dat veel zaken die we hier bespreken, je toegang verschaffen tot het onbewuste van een vrouw. Zodra zij dat doorheeft werpt ze onmiddellijk een blokkade op.' Gorny and Ewijk, *MasterFlirt*, 3.
28. Pettit, *The Science of Deception.*
29. See also Coon, 'Testing the Limits of Sense and Science. American Experimental Psychologists Combat Spiritualism, 1880–1920.' About the history of the lie detector see Geoff Bunn's brilliant *The Truth Machine.*
30. For a history of deception in social psychology see James Korn's *Illusions of Reality.*
31. Pettit, *The Science of Deception*, 218.
32. Rutherford, *Beyond the Box*; Baistow, 'Behavioural Approaches and the Cultivation of Competence'.
33. Latour, '*The Prince* for Machines as Well as for Machinations'.
34. Ibid., 21.
35. Ibid.
36. Ibid., 22.
37. Ibid., 36.
38. Ibid.
39. Latour, *Reassembling the Social*, 68.
40. Latour, 'When Things Strike Back', 113.

41. Latour, 'Where Are the Missing Masses? The Sociology of a Few Mundane Artifacts'.
42. For an early discussion of the materialist turn, see Pels, Hetherington and Vandenberghe, 'The Status of the Object Performances, Mediations, and Techniques', and the special issue of which their essay is the introduction.
43. Latour, *The Pasteurization of France*, 9.
44. Ibid., 10.
45. Ibid., 158.
46. Latour, 'When Things Strike Back', 115.
47. Following the work of Isabelle Stengers and Vinciane Despret.
48. Latour, 'When Things Strike Back', 116; emphasis in original.
49. Ibid.
50. Machiavelli, *The Prince*, 21.
51. Ibid., 59.
52. Ibid., 37.
53. Ibid., 39.
54. Ibid., 59.
55. Paul, 'The Use of Kairos in Renaissance Political Philosophy'.
56. Machiavelli, *The Prince*, 62.
57. Ibid., 28.
58. Ibid., 9.
59. Ibid., 66. Machiavelli notes that 'the people are now more powerful' (than the soldiers), and thus it is more important to please them than the soldiers, but he doesn't develop it further. Ibid., 88.
60. Ibid., 74.
61. Ibid., 62.
62. Ibid., 52.
63. Russell Price, the translator, notes here that this apparently false argument can be saved by assuming Machiavelli means 'good order' when he writes 'good laws': to be able to raise an army a state must be well-ordered, even though it may lack an adequate or just legal system.
64. Foucault, *Discipline and Punish.*
65. Rose, *Inventing Our Selves*, 120.
66. Ibid., 121.
67. Rose, 'Engineering the Human Soul', 367.
68. Foucault, 'Technologies of the Self', 19.
69. Karlsen and Villadsen, 'Who Should Do the Talking?'
70. Ibid., 347.
71. Ibid., 360.
72. See also my own analysis of the psychological test as a tool that forces spontaneity: Derksen, 'Discipline, Subjectivity and Personality'.
73. Lezaun, 'A Market of Opinions', 130.
74. See also my discussion of Lezaun's study in Derksen, 'People as Scientific Instruments'.
75. 'A Market of Opinions', 136.
76. Ibid., 138.
77. Ibid., 141.

78. Ibid., 135.
79. Callon and Law, 'On Qualculation, Agency, and Otherness'.
80. For this point Callon and Law build on Callon and Muniesa, 'Economic Markets as Calculative Collective Devices'. They borrow the term 'qualculation' from Cochoy to indicate that processes of calculation are not necessarily quantitative. I agree but will stick with 'calculation'.
81. Callon and Law, 'On Qualculation, Agency, and Otherness', 723.
82. Ibid.
83. Rose, 'Engineering the Human Soul', 368.
84. Rose, *Inventing Our Selves*, 121.
85. In his article 'Engineering the Human Soul' the human technologies are the engineers themselves. Rose, 'Engineering the Human Soul', 368.
86. 'Technology, Testing Text: Clinical Budgeting in the U.K. National Health Service'.
87. Ibid., 266.
88. Anne Beaulieu and I first proposed this approach to 'social technology' as an invitation to study differences in Derksen and Beaulieu, 'Social Technology'. See also Derksen, Vikkelsø and Beaulieu, 'Social Technologies'.

3 Scientific Management and the Human Factor

1. Duncan, 'Efficiency – Real, Unreal, and Brutal', 383.
2. Frey, 'The Relation of Scientific Management to Labor', 300.
3. Baker, 'The Early History of the Term "Social Science"'.
4. Condorcet, 'Outline of an Historical View of the Progress of the Human Mind', 276, 273.
5. Latour, 'When Things Strike Back', 118. Latour builds on work of Zygmunt Bauman.
6. Ibid.
7. Hughes, *American Genesis*, 187.
8. Benjamin and Baker, *From Séance to Science*, 130.
9. Benjafield, *A History of Psychology*, 138.
10. Taylor, 'A Piece-Rate System'.
11. Kaufman, *Managing the Human Factor*, 58.
12. Nelson, *Frederick W. Taylor and the Rise of Scientific Management*, 8.
13. Ibid., 16.
14. Ibid., 169.
15. Nadworny, 'Scientific Management and the Unions, 1900–1932', 24.
16. Gompers, 'Editorial', 116.
17. Gompers, 'The Miracles of Efficiency', 277.
18. Evans, 'What Is to Become of the Machinist?'
19. Marx, *Capital*, 461–62. Kang, in *Sublime Dreams of Living Machines*, discusses the trope of the living machine in the industrial age and shows that it had positive as well as negative connotations.
20. '[L]'ouvrier doit devenir, il est devenu, un automate.' Cited in Bloemen, *Scientific Management in Nederland, 1900–1930*, 55.

21. Cited in Friedmann, *Industrial Society; the Emergence of the Human Problems of Automation*, 43.
22. Van Ginneken, *Zielkunde en Taylor-systeem*, 8.
23. Frey, 'The Relation of Scientific Management to Labor', 297.
24. US Congress. House. Committee on Labor, *Investigation of Taylor System of Shop Management. Hearings before the Committee on Labor of the House of Representatives, Sixty-Second Congress*, 24.
25. Ibid., 27.
26. Gompers, 'Speeding Up', 525.
27. Duncan, 'Efficiency – Real, Unreal, and Brutal', 383.
28. '[L]a personnalité, l'intelligence, les désirs mêmes de l'ouvrier sont effacés', cited in Bloemen, *Scientific Management in Nederland, 1900–1930*, 54.
29. Van Ginneken, 'Zielkunde en Taylor-systeem', 22.
30. Ibid., 32.
31. For the enthusiastic implementation of Taylorism and Fordism in Soviet Russia, see Hughes, *American Genesis*, chapter 6. He quotes the poet and union leader Alexei Gastev on p. 258: 'Many find it repugnant that we want to deal with human beings as with a screw, a nut, a machine. But we must undertake this as fearlessly as we accept the growth of trees and the expansion of the railway network.'
32. Weil, 'Prospects. Are We Heading for a Proletarian Revolution?', 10.
33. US Congress. House. Committee on Labor, *The Taylor and Other Systems of Shop Management. Hearings before Special Committee of the House of Representatives to Investigate the Taylor and Other Systems of Shop Management*, 1222.
34. US Congress. House. Committee on Labor, *Investigation of Taylor System of Shop Management. Hearings before the Committee on Labor of the House of Representatives, Sixty-Second Congress*, 5.
35. US Congress. House. Committee on Labor, *The Taylor and Other Systems of Shop Management. Hearings before Special Committee of the House of Representatives to Investigate the Taylor and Other Systems of Shop Management.*
36. Ibid., 101.
37. Taylor, 'The Principles of Scientific Management'.
38. In all likelihood, it was written together with, or even entirely by, his wife Lillian Moller Gilbreth, about whom there more follows. Lancaster, *Making Time*, 118.
39. Gilbreth, *Primer of Scientific Management*, 49.
40. Ibid., 50.
41. Ibid.
42. Taylor, 'Testimony before the Special House Committee', 33.
43. Ibid., 7.
44. Ibid., 27.
45. Ibid., 49.
46. Ibid., 7.
47. Ibid., 30.
48. Cited in Kanigel, *The One Best Way*, 477.

49. Ibid., 283. Kanigel sees the influence of Taylor's Quaker background in his social idealism.
50. Taylor, 'Shop Management', 30.
51. Ibid., 184.
52. Ibid., 27. See Oldenziel, *Making Technology Masculine*, chapter 2, for the symbolic meaning of shop floor manliness for engineers of Taylor's generation. As engineers became managers, it became all the more important to mark their physical prowess and assert their authority by positioning themselves as one of the boys yet also above the boys.
53. Taylor, 'Shop Management', 143.
54. Taylor, 'Testimony before the Special House Committee', 45.
55. Taylor, 'Shop Management', 176.
56. Taylor, 'Testimony before the Special House Committee', 264.
57. Taylor, 'Shop Management', 182.
58. Taylor recounts the episode in 'The Principles of Scientific Management', 44.
59. Taylor, 'The Principles of Scientific Management', 6.
60. Nelson, *Frederick W. Taylor and the Rise of Scientific Management*. On the other hand, the manly camaraderie that Taylor sketches as part of the essence of scientific management did reflect the way he actually related to the workers.
61. US Congress. House. Committee on Labor, *The Taylor and Other Systems of Shop Management. Hearings before Special Committee of the House of Representatives to Investigate the Taylor and Other Systems of Shop Management*, 3 [Report]. The report was published in a separately numbered volume.
62. US Congress. House. Committee on Labor, *The Taylor and Other Systems of Shop Management. Hearings before Special Committee of the House of Representatives to Investigate the Taylor and Other Systems of Shop Management*, 4 [Report].
63. Kaufman, *Managing the Human Factor*, 69.
64. Anonymous, 'The Mistakes of the Efficiency Men'.
65. Peck, 'Systematic versus Scientific Management'.
66. Frey, 'The Relation of Scientific Management to Labor'.
67. Scott, *Increasing Human Efficiency in Business*, 3.
68. Ibid., 255.
69. Münsterberg, *Psychology and Industrial Efficiency*, 50.
70. Ibid., 56.
71. Ibid., 53.
72. Quoted in Lancaster, *Making Time*, 120.
73. Ibid., 95.
74. Ibid., 107.
75. Ibid., 149.
76. Ibid., 111.
77. Ibid., 115.
78. She eventually went to Brown and earned her doctorate in 1915 for a thesis on applying efficiency methods in schools.

79. Lancaster, *Making Time*, 125.
80. Gilbreth, *The Psychology of Management*, 20.
81. Ibid., 19.
82. Ibid., 32.
83. Ibid., 36–37. Among the more outlandish arguments in the book is Moller Gilbreth's connection of output records and genius: 'The outcome of the records, and their related elements in other branches of Scientific Management, is to arouse interest. Interest arouses abnormally concentrated attention, and this in turn is the cause of genius. This again answers the argument of those who claim that Scientific Management kills individuality and turns the worker into a machine.' Ibid., 43.
84. Ibid., 40.
85. 'Far from making machines out of the men, standardization causes a mental state that leads to invention, for the reason that the worker's brain is in most intimate contact with the work, and yet has not been unnecessarily fatigued by the work itself.' Gilbreth, *The Psychology of Management*, 180.
86. Ibid., 177.
87. Ibid., 2.
88. Ibid., 2.
89. Ibid., 71.
90. Ibid., 221.
91. Ibid., 238.
92. Nelson, 'Scientific Management, Systematic Management, and Labor, 1880–1915'.
93. Cooke, Gompers and Miller, *Labor, Management and Production*, 7.
94. See e.g. Friedmann, *Industrial Society; the Emergence of the Human Problems of Automation*, 45 ('physiology and psychology ... prepared the way for developing a science of industrial labor, aiming at gradually introducing into the workshops and extending there the consideration of the human factor'); and recently Van de Water, 'Psychology's Entrepreneurs and the Marketing of Industrial Psychology'.
95. Schachter, 'Frederick Winslow Taylor and the Idea of Worker Participation'.
96. Taylor, 'The Principles of Scientific Management', 96.
97. See Gillespie, *Manufacturing Knowledge*, 162 ff., for a description of this study.
98. Taylor, 'The Principles of Scientific Management', 119.
99. Gilbreth, *The Psychology of Management*, 179.
100. Lancaster, *Making Time*, 153–54.
101. Brown, *The Definition of a Profession*.
102. Münsterberg, *On the Witness Stand: Essays in Psychology and Crime*.
103. Brown, *The Definition of a Profession*, 108.
104. Myers, *Present-Day Applications of Psychology, with Special Reference to Industry, Education and Nervous Breakdown*.
105. Myers, 'The Efficiency Engineer and the Industrial Psychologist'.
106. Drever, 'The Human Factor in Industrial Relations'.

107. And psychoanalysis: Myers, *Mind and Work*, 166ff., employed the concept of defense mechanisms to shed light on the psychology of industrial unrest.
108. US Congress. House. Committee on Labor, *The Taylor and Other Systems of Shop Management. Hearings before Special Committee of the House of Representatives to Investigate the Taylor and Other Systems of Shop Management*, 18.
109. Taylor, 'Testimony before the Special House Committee', 59. See also Gilbreth, *The Psychology of Management*, 217, who refers to Gantt in support.
110. Cooke, Gompers and Miller, *Labor, Management and Production*, iii.
111. Richard Gillespie, in his history of the Hawthorne experiments, describes the rise of personnel management as a response to the growing power of workers and their unions and the resultant need to win their loyalty and commitment. Gillespie, *Manufacturing Knowledge*. The motives and attitudes of employees became a central concern of management.
112. A. J. Todd, in Cooke, Gompers and Miller, *Labor, Management and Production*, 31.
113. Gompers, in ibid., xii.
114. Follett, 'The Giving of Orders'.
115. Her papers are collected in *Dynamic Administration*, edited by Elliot Fox and L. Urwick. They also give an excellent introduction to her life and work.
116. Follett, 'The Giving of Orders', 24.
117. Ibid., 33.
118. Ibid., 32.
119. Herein lies one of the greatest contributions of scientific management, she adds: 'it tends to depersonalize orders'. Ibid., 30.
120. Ibid., 33.
121. Ibid., 35.
122. Cawley and Chaloupka, in 'American Governmentality', argue that Follett refined and elaborated tactics of management that Taylor had pioneered.
123. Follett, 'The Process of Control', 169.
124. Gilbreth, *Primer of Scientific Management*, 73.

4 'Social Technology'

1. Small, 'Seminar Notes', 113.
2. Ibid., 131.
3. Ibid.
4. Oldenziel, *Making Technology Masculine*. The classic exploration of keywords, including 'technology', is Williams, *Keywords. A Vocabulary of Culture and Society*.
5. Popper, 'The Poverty of Historicism, II. A Criticism of Historicist Methods'.
6. Merton, *Social Theory and Social Structure*.
7. Cartwright, 'Achieving Change in People'.
8. Helmer, Brown and Gordon, *Social Technology*.
9. The search terms used were 'social technology' and 'social engineering', case insensitive.

10. Search performed on October 17, 2013.
11. Marx, 'Technology', 563.
12. For the history of American sociology I build in particular on Haskell, *The Emergence of Professional Social Science*; Bulmer, *The Chicago School of Sociology*, Bannister, *Sociology and Scientism*; Ross, *The Origins of American Social Science*; Wilde, *Discipline en legende*; and Smith, *Social Science in the Crucible*.
13. Bulmer, *The Chicago School of Sociology*, 33.
14. For a history of the boundary work of early American and German sociologists, see Wilde, *Discipline en legende*.
15. Small, 'Fifty Years of Sociology in the United States (1865–1915)', 726.
16. Small and Vincent, *An Introduction to the Study of Society*, 74.
17. Bulmer, *The Chicago School of Sociology*, 33.
18. Schmoller, in Boese, *Geschichte des Vereins für Sozialpolitik*.
19. '[W]ir halten die Gesellschaft für die normalste and gesündeste, die eine Stufenleiter verschiedener Existenzen, aber mit leichtem Übergang von einer Sprosse zur anderen darstellt'. Schmoller, in ibid., 9.
20. 'grossartigste sittliche Institut der Erziehung des Menschengeschlechts'. Schmoller, in ibid., 8.
21. Bannister, *Sociology and Scientism*.
22. Ross, *The Origins of American Social Science*, 126–27.
23. Small and Vincent, *An Introduction to the Study of Society*, 41.
24. Small, 'General Sociology', 213.
25. Frison, 'Some German and Austrian Ideas on "Technologie" and "Technik"', 120.
26. Post, *Arbeit statt Almosen*.
27. Natorp, *Sozialpädagogik*.
28. Ofner, *Studien sozialer Jurisprudenz*.
29. Ibid., 25.
30. Ibid., 28.
31. Ibid., 26.
32. In 'Platos Staat und die Idee der Sozialpädagogik', which originally appeared in the *Archiv für soziale Gesetzgebung und Statistik*, vol. 8, 1895, and was reprinted in Natorp, *Gesammelte Abhandlungen zur Sozialpädagogik*.
33. For a biography of Henderson, see Abbott, 'Pragmatic Sociology and the Public Sphere'.
34. Small, 'Fifty Years of Sociology in the United States (1865–1915)', 770.
35. Abbott, 'Pragmatic Sociology and the Public Sphere'; Odum, *American Sociology*.
36. Odum, *American Sociology*, 397.
37. Henderson, 'The Scope of Social Technology'.
38. Vincent writes that 'social technology' was proposed by Henderson. Vincent, 'Varieties of Sociology', 9. He might be referring to the 1901 article here.
39. Henderson, 'The Scope of Social Technology', 467.
40. Ibid., 468.
41. Ibid.
42. Ibid., 481.

43. Ibid., 474.
44. Ibid.
45. Ibid., 476.
46. Ibid., 477.
47. Ibid., 484.
48. Ibid., 471.
49. Henderson, 'Applied Sociology (or Social Technology)', 217.
50. Ibid., 216. A similar conception of social technology was later formulated by Dwight Sanderson: 'As I see it any social technology must involve the application of many sciences, not only social but physical and biological.' Sanderson, 'Discussion of Sociometry', 215.
51. Henderson, 'Applied Sociology (or Social Technology)', 221.
52. Tolman, 'The Study of Sociology'.
53. Bernard, 'The Teaching of Sociology in the United States'.
54. Ibid., 164.
55. Ibid., 165.
56. The lack of agreement about the classification of sociological subjects was still a problem for the 'Survey of Sociology in American colleges' conducted by Kennedy and Kennedy, 'Sociology in American Colleges'.
57. Haskell, *The Emergence of Professional Social Science*, 206.
58. Giddings, 'Relation of Sociology to Scientific Studies', 145. About this address see the analysis by Haskell, *The Emergence of Professional Social Science*. The relation between Giddings and Small and their respective versions of sociology is analysed by Bannister, *Sociology and Scientism;* and Ross, *The Origins of American Social Science*.
59. Giddings, 'Sociology as a University Study', 650–51.
60. Giddings, *The Principles of Sociology*, 130.
61. Ibid., 131.
62. Ross, *The Origins of American Social Science*, 127.
63. Henderson, 'The Scope of Social Technology', 475.
64. Ibid.
65. Steiner received his PhD in 1915 from Small and, according to Odum, *American Sociology*, 375, was greatly influenced by Small and Henderson. Like Small, Vincent and Henderson, Steiner was trained in the ministry. Odum mentions one other (associate) professor of social technology: Stuart Queen at Goucher College, 1919–1920. Queen was also a Chicago PhD.
66. Steiner, 'The Limitations of a Conceptual Approach to the Applications of Sociology to Social Work', 501.
67. Ibid., 502.
68. Taylor, 'The Social Survey and the Science of Sociology'.
69. Bogardus, *A History of Social Thought*.
70. Bogardus, *Introduction to Sociology*.
71. Small, 'The Scope of Sociology. IX. Premises of Practical Sociology', 26.
72. Small, *General Sociology. An Exposition of the Main Development in Sociological Theory from Spencer to Ratzenhofer*, 34.
73. Ibid., 317. In both cases Small added that social technology 'bears the same relation to fundamental sociology that all physical technology bears to the

underlying physical sciences.' Small, 'The Scope of Sociology. IX. Premises of Practical Sociology', 26.

74. Small, 'Fifty Years of Sociology in the United States (1865–1915)', 828.
75. A sample from the *American Journal of Sociology*: Blackman, 'Reasonable Department of Sociology for Colleges and Universities'; Dewey, 'An Application of Statistical Method'; Hayes, 'Sociology as Ethics'; Henderson, 'Definition of a Social Policy Relating to the Dependent Group'; Lind, 'Some Ecological Patterns of Community Disorganization in Honolulu'; Riley, 'Sociology and Social Surveys'; Small, 'The Scope of Sociology. VI. Some Incidents of Association'; Small, 'The Scope of Sociology. VII. Classification of Associations'; Small, 'Points of Agreement among Sociologists'; Small, 'Fifty Years of Sociology in the United States (1865–1915)'.
76. Steiner, *Education for Social Work*.
77. Ellwood, 'Sociology: Its Problems and Its Relations', 326.
78. Burgess and Park, *Introduction to the Science of Sociology*, 43.
79. Breslau, 'The Scientific Appropriation of Social Research', 428.
80. Throop and Ward, 'Organization of Academic Foci within the "Department of Social Science"', write: 'During the period when Small was Head Professor (1892–1925), there were relatively few changes. General Sociology at first included Social Philosophy and Social Institutions. The two sub foci pair separated and enlarged into "Social Philosophy" (taught by Small) and "Social Institutions" (taught by Henderson, later joined by Vincent). Institutions was renamed "Social Technology" in 1901, remaining so until Small's last year.'
81. Breslau, 'The Scientific Appropriation of Social Research', 432.
82. Park, quoted in ibid., 438.
83. See, in particular, Bannister, *Sociology and Scientism*; Ross, *The Origins of American Social Science*; and Smith, *Social Science in the Crucible*.
84. Among sociologists, at least. Henderson's funeral was attended by thousands of people. Abbott, 'Pragmatic Sociology and the Public Sphere'. The case can also be made that the empirical work of the Chicago School owed much to the influence of Henderson, whose courses had required students to do small empirical investigations in social problems. Bulmer, *The Chicago School of Sociology*; Wallace, 'Starting a Department and Getting It Under Way'.
85. Small, in *Publication of the American Sociological Society*, 8:159.
86. Ibid.
87. Ibid., 8:160.
88. Park, in ibid., 8:167.
89. Ibid., 8:168.
90. Burgess and Park, *Introduction to the Science of Sociology*, 46.
91. Ibid., 339.
92. Jordan, *Machine-Age Ideology*, 39. See also Etzemüller, *Die Ordnung der Moderne*, for a collection of essays about social engineering in the twentieth century. Dennis Bryson has extensively studied the involvement of American social scientists and psychologists in social and human engineering; see, for example, Bryson, 'Lawrence K. Frank, Knowledge, and the Production of the "Social"'.

93. For a history of the technocrats see Jordan, *Machine-Age Ideology;* and particularly Akin, *Technocracy and the American Dream.*
94. Jordan, *Machine-Age Ideology.*
95. Ogburn, 'The Folkways of a Scientific Sociology'.
96. Henderson, *Social Elements. Institutions, Character, Progress.*
97. Ibid., 94.
98. In an 1895 review of Amos Warner's *American Charities*, Henderson briefly discussed Warner's distinction between science and art, agreeing with Warner that the art of philanthropy needs to be informed by science in order to make progress. Henderson, 'Review'. Warner included medicine and nursing in the arts.
99. Oldenziel, *Making Technology Masculine.*
100. Marx, 'Technology'; Williams, *Keywords. A Vocabulary of Culture and Society.*
101. Schatzberg, 'Technik Comes to America', 493–94.
102. See also Kline, 'Construing "Technology" as "Applied Science"'. Williams, *Keywords. A Vocabulary of Culture and Society*, noted the same shift in meaning.
103. Schatzberg, 'Technik Comes to America', 494.
104. Ibid., 495.
105. Ibid., 500.
106. In fact, Veblen's use of 'technology' 'combined the idea of technology as Technik with the original meaning of technology as a field of knowledge'. Ibid., 503., but this nuance was not picked up by his readers.
107. Akin, *Technocracy and the American Dream.*
108. Veblen, 'The Engineers and the Price System'.
109. Ibid., 140.
110. Maier, 'Between Taylorism and Technocracy'.
111. Ogburn, 'Technology and Sociology'.
112. This description, twenty-four years later, of his state of mind in 1892 reveals his ambivalence regarding science and reform: 'I had also been growing more and more dissatisfied with the ways in which the men were proceeding who had the impulses which I most respected about efforts to control the facts of society in the interest of social betterment.' Small, 'Fifty Years of Sociology in the United States (1865–1915)', 768.
113. It is interesting to note that Veblen's introduction of 'technology', which was responsible for that shift, came three years after Small's coining of 'social technology'. Both worked at the University of Chicago and knew each other well. It is possible (although I have found no evidence for it) that Veblen's 'technology' was inspired by Small's 'social technology'.
114. About the relation between sociology and social work see Breslau, 'The Scientific Appropriation of Social Research'; Jordan, *Machine-Age Ideology*; and in particular Lubove, *The Professional Altruist.*
115. Lubove, *The Professional Altruist*, 143.
116. Eliot used quotes he had gathered at various conferences.
117. Eliot, 'The Social Worker's Criticisms of Undergraduate Sociology'.
118. Lubove, *The Professional Altruist.*

119. Veblen, in 'Some Neglected Points in the Theory of Socialism', had already used the term 'social engineering' in 1891, in one of his first publications, a commentary on Herbert Spencer's essay 'From Freedom to Bondage'.
120. Jordan, *Machine-Age Ideology*, 3.
121. About the masculinisation of engineering and technology see Oldenziel, *Making Technology Masculine.*
122. Ibid., 46.
123. Kline, 'Construing "Technology" as "Applied Science"'.
124. Oldenziel, *Making Technology Masculine*, 46.
125. Bulmer and Bulmer, 'Philanthropy and Social Science in the 1920s', 385.
126. Ibid., 387.
127. Ibid., 385.
128. Bernard, 'Standards of Living and Planes of Living', 193.
129. Ibid., 196.
130. Ibid.
131. Bryson, 'Lawrence K. Frank, Knowledge, and the Production of the "Social"'.
132. Jordan, *Machine-Age Ideology*, 190, 276.
133. Ibid.
134. Smith, *Social Science in the Crucible*, 5.
135. Marklund, 'Begriffsgeschichte and Übergriffsgeschichte in the History of Social Engineering'.
136. Oldenziel, *Making Technology Masculine.*
137. Lubove, *The Professional Altruist*, 6; Wilde, *Discipline en legende.*
138. Ogburn, 'The Folkways of a Scientific Sociology', 2.
139. Etzemüller, '*Social engineering* als Verhaltenslehre des kühlen Kopfes', 35.

5 Dale Carnegie and the Fine Art of Dealing with People

1. Carnegie, *How to Win Friends and Influence People*, 135.
2. Ibid., 23.
3. Ibid., vii.
4. Carnegie changed the name of the course several times. His YMCA course started as the YMCA Course in Public Speaking. It was later also known as the Carnegie Course in Public Speaking and Human Relations, the Dale Carnegie Course in Public Speaking and Human Relations, the Dale Carnegie Course in Effective Speaking and Human Relations, and, after the success of *How to Win* . . ., the Dale Carnegie Leadership Course. Informally it was usually called the Carnegie Course.
5. Guinn, *Manson.*
6. Carnegie, *How to Win Friends and Influence People*, 20.
7. *New York Times*, December 7, 1936.
8. *New York Times*, January 10, 1937.
9. *New York Times*, February 1 and 3, 1937.
10. *New York Times*, February 25, 1937.
11. *New York Times*, April 3, 1938.

12. See for this type of analysis: Cain, *Quiet*; Ehrenreich, *Bright-Sided*; McGee, *Self-Help, Inc.*; Meyer, *The Positive Thinkers*; and Susman, *Culture as History*.
13. Susman, *Culture as History*, 200.
14. For example, the *New York Times* of March 29, 1938, announces that Dale Carnegie, one of the authors in the 'currently popular "How to –" category', will speak at a luncheon of the Fashion Group, together with Gelett Burgess, author of *How to Look Eleven Years Younger*; other authors of 'How to –' titles were also present.
15. Susman, *Culture as History*, 166.
16. The Library of Congress lists *How to Win* ... among the 'books that shaped America': www.loc.gov/bookfest/books-that-shaped-america/.
17. In Chapter 1 of this book, Rick van Baaren similarly warned against using techniques as tricks.
18. See Steven Watts's *Self-Help Messiah* for an excellent biography of Carnegie.
19. Watts, *Self-Help Messiah*.
20. Sproule, 'Inventing Public Speaking', 578.
21. Esenwein and Carnagey, *The Art of Public Speaking*, 24.
22. Ibid., 49.
23. Ibid.
24. Ibid., 113.
25. Ibid., 161.
26. Ibid., 6.
27. Ibid., 56.
28. Kuna, 'The Concept of Suggestion in the Early History of Advertising Psychology'.
29. James, *The Principles of Psychology*, 1134.
30. Recently Kevin Dutton, in his book *Flipnosis*, has again linked suggestion, hypnosis and persuasion.
31. Münsterberg, 'The Market and Psychology', 153.
32. Ibid., 170.
33. Ibid.
34. Ibid., 170–71. See Carnegie *Public Speaking; a Practical Course for Business Men*, 388.
35. Münsterberg, *Psychology and Industrial Efficiency*, 294.
36. Ibid., 297.
37. Ibid., 299.
38. Scott, *Influencing Men in Business; the Psychology of Argument and Suggestion*, 9.
39. Ibid., 35.
40. Ibid., 83.
41. Ibid., 93.
42. Esenwein and Carnagey, *The Art of Public Speaking*, 8.
43. That either he or Esenwein had read it is certain: their chapter on influencing the crowd is a scrambled, rewritten version of a chapter in Scott's book.
44. Scott, *The Psychology of Public Speaking, Adapted for Use as a Text Book in High Schools and Colleges*.
45. Ibid., 66.

46. Ibid.; Carnegie, *How to Win Friends and Influence People*, 271.
47. Ramage, *Twentieth-Century American Success Rhetoric*, 104.
48. Carnagey, *Public Speaking*, 9.
49. Ibid., 17.
50. Ibid., 30–31.
51. Ibid., 23.
52. Ibid., 22.
53. Ibid., 23.
54. Ibid., 8.
55. Ibid., 3.
56. Ibid., 4–5.
57. Ibid., 4.
58. Ibid., 18.
59. Ibid., 16.
60. Ibid., 4.
61. Carnegie, *Public Speaking; a Practical Course for Business Men*. Watts, *Self-Help Messiah*, 160, dates the change to sometime between November 1924 and October 1925.
62. For example, the *New York Times* of December 13, 1936, featured advertisements for public speaking trainer Walter O. Robinson and for the Carnegie School for Expression, both at Carnegie Hall.
63. Latour, 'Visualization and Cognition: Thinking with Eyes and Hands'.
64. Carnegie, *Public Speaking; a Practical Course for Business Men*, 174ff.
65. Ibid., 18.
66. Ibid., 197.
67. Ibid., 242.
68. Ibid., 204.
69. Ibid., 242.
70. Ibid., 245.
71. Ibid., 201.
72. Ibid., 225.
73. Ibid., 238.
74. Ibid., 202.
75. Ibid., 385.
76. Ibid., 296.
77. Carnegie's anecdote was a gross simplification of a long and bloody labour conflict that is hardly done justice with the perfunctory '[s]hooting had taken place; there had been bloodshed'. Ibid., 295. In the Ludlow Massacre, the most gruesome event in this conflict, around twenty people, including women and children, burnt to death when the tent in which they were hiding was set alight. Rockefeller's shock over the violence prompted him to intervene, but the reaction to his proposals was far from unanimously positive.
78. The same speech – 'Friends, Romans, countrymen, lend me your ears' – serves as an illustration of indirect suggestion in Scott's *The Psychology of Public Speaking, Adapted for Use as a Text Book in High Schools and Colleges*, 107, 167 – another indication perhaps that Carnegie had read Scott's book.

79. Carnegie, *Public Speaking; a Practical Course for Business Men*, 424.
80. Carnagey, *Public Speaking*, 18.
81. However, the implied audience of Carnegie's books were men. Only with *How to Win Friends and Influence People* did the implied audience come to include women.
82. Carnegie, *Public Speaking; a Practical Course for Business Men*, 451.
83. Watts, *Self-Help Messiah*, 235.
84. Overstreet, *Influencing Human Behavior*, 1.
85. Ibid.
86. Ibid., 3.
87. Ibid., 276.
88. Ibid., 5.
89. Ibid., 49.
90. Ibid., 46.
91. Ibid., 47–48.
92. Ibid., 171.
93. Carnegie, *How to Win Friends and Influence People*, 20.
94. Carnegie, *Comment se faire des amis.*
95. Carnegie, *How to Win Friends and Influence People*, 42.
96. Ibid., 43.
97. Ibid.
98. Ibid., 55.
99. Ibid., 58.
100. Ibid., 83.
101. Ibid., 147.
102. Ibid., 243. The distinction between these two topics – influencing people and effecting long-term change in them – was another element inspired by Overstreet's book.
103. Schwab was Andrew Carnegie's CEO, the best paid manager of his day and one of Dale Carnegie's heroes.
104. Carnegie, *How to Win Friends and Influence People*, 158.
105. Ibid., 57.
106. Ibid., 221.
107. Ibid., 42.
108. Ibid., 207.
109. Ibid., 213.
110. Ibid., 53.
111. Ibid., 54.
112. Ibid., 40.
113. Ibid., 95.
114. Ibid.
115. Ibid., 145.
116. Ramage, *Twentieth-Century American Success Rhetoric*, 109.
117. Carnegie, *How to Win Friends and Influence People*, 41.
118. Ibid., 100.
119. *Public Speaking* has a similar introduction by Thomas.
120. Carnegie, *How to Win Friends and Influence People*, 8.

121. Ibid., 22.
122. *New York Times*, November 27, 1938.
123. Alexander, 'The Green Pencil', 50.
124. Tressler, *How to Lose Friends and Alienate People*, 34.
125. *New York Times*, March 22, 1938.
126. 'Soft Answers'.
127. Thurber, 'The Voice with the Smile'.
128. 'Pastor Raps Best Seller. Asserts Dale Carnegie's Book Is a Mockery of True Friendship'.
129. Lewis, 'Car-Yes-Man'.
130. Ibid.
131. Lewis, 'One Man Revolution'. Posthumously adding insult to injury, Pulitzer critic Jonathan Yardley, in *American Heritage* magazine, ranked *How to Win* ... in the top ten of most influential American books of the twentieth century but placed two of Sinclair Lewis's books on his 'also-ran' list. 'Pulitzer Critic Rates Top Ten Culture Books'.
132. McGee, *Self-Help, Inc.*, 36.
133. Ehrenreich, *Bright-Sided*, 52.
134. Meyer, *The Positive Thinkers*, 169.
135. Ramage, *Twentieth-Century American Success Rhetoric*, 114.
136. Ibid., 115.
137. Lewis, 'One Man Revolution'.
138. Saroyan, *The Trouble with Tigers*, 277.
139. Ibid., 278.
140. Ibid., 279.
141. Ibid., 281.
142. Ibid., 283.

6 Karl Popper's Social Technology and the Personal Element

1. Jordan, *Machine-Age Ideology*.
2. Ibid., 285–86.
3. Bernard, 'The Place of the Social Sciences in Modern Education', 1935, 52.
4. Ibid., 52.
5. Ibid., 52–53.
6. Ibid., 53.
7. Ibid., 55.
8. Bernard, 'Some Dynamics of Present World Relations', 109.
9. Ibid., 110.
10. Ibid., 106.
11. Hacohen, 'Karl Popper, the Vienna Circle, and Red Vienna'; Hacohen, *Karl Popper, the Formative Years, 1902–1945*; Popper, *Unended Quest*.
12. For a recent demonstration that very few scientific studies conform to falsificationist principles, see Hansson, 'Falsificationism Falsified'. Bunge, 'The Seven Pillars of Popper's Social Philosophy', presents an incisive critique of Popper's political philosophy.

13. See Hacohen, *Karl Popper, the Formative Years, 1902–1945*, 353–54. Later, Popper prepared a revised and extended book version, which appeared in 1957.
14. Popper, 'The Poverty of Historicism, II. A Criticism of Historicist Methods', 120.
15. Popper, 'The Poverty of Historicism, I', 99.
16. Popper, 'The Poverty of Historicism, II. A Criticism of Historicist Methods', 119.
17. It has been noted, for example by Ray, 'Critical Theory and Positivism'; Bunge 'The Seven Pillars of Popper's Social Philosophy'; and Pickel 'Between Social Science and Social Technology'; that the relationship between social science and social technology is not entirely clear in Popper's work.
18. Popper, 'The Poverty of Historicism, II. A Criticism of Historicist Methods', 120.
19. Ibid.
20. Ibid., 121.
21. Ibid., 122.
22. Ibid., 123.
23. Ibid., 122.
24. Simkin, *Popper's Views on Natural and Social Science*, 184.
25. Hacohen, 'Karl Popper, the Vienna Circle, and Red Vienna', 714.
26. Ibid., 715.
27. Ibid., 719.
28. Popper, *The Open Society and Its Enemies. Volume II. The High Tide of Prohecy: Hegel, Marx, and the Aftermath*, 210–11.
29. Hacohen, *Karl Popper, the Formative Years, 1902–1945*, 370–71. Nevertheless, Hacohen (in ibid., 384) mentions 'social engineering' among the terms that Popper 'coined or defined', which is clearly incorrect. In the fifth edition of *The Open Society*, Popper himself traces the history of the term 'social engineering' back to a book by Roscoe Pound from 1922. There are older sources, however, such as William Tolman's *Social Engineering* from 1909.
30. Ibid., 373.
31. Ibid., 374.
32. Popper, 'The Poverty of Historicism, II. A Criticism of Historicist Methods', 123.
33. Ibid.
34. Ibid., note 5.
35. Ibid., 124.
36. Ibid., 128.
37. Ibid.
38. Popper, 'The Poverty of Historicism, III', 88.
39. Ibid.
40. Ibid.
41. Ibid.
42. Ibid.

43. Ibid.
44. Popper, *The Open Society and Its Enemies. Volume II. The High Tide of Prophecy: Hegel, Marx, and the Aftermath*, 93.
45. Ibid., 97.
46. Popper, *The Open Society and Its Enemies. Volume I: The Spell of Plato*, 1:135.
47. Ibid., 1:133. Popper thought the best answer was a two-party system, as in Britain and the USA. Bunge, 'The Seven Pillars of Popper's Social Philosophy', 548.
48. Popper, *The Open Society and Its Enemies. Volume I: The Spell of Plato*, 1:134.
49. Ibid., 1:136.
50. Bunge, 'The Seven Pillars of Popper's Social Philosophy'.

7 Tactful Leadership

1. See Marrow, *The Practical Theorist*, for a biography of Lewin and Cravens, *Before Head Start*, for Lewin's years at the Iowa Child Welfare Research Station.
2. John, Eckardt and Hiebsch, 'Kurt Lewin's Early Intentions (Dedicated to His 100th Birthday)'.
3. Lewin, *Die Sozialisierung des Taylorsystems*.
4. 'Wer einmal Versuchsperson gewesen ist, wird bestätigen, daß es sich in der Regel um nicht uninteressante Vorgänge handelt, bei denen man nicht selten auch etwas für seine persönliche Arbeitsweise lernen kann.' Ibid., 4:19.
5. John, Eckardt and Hiebsch, 'Kurt Lewin's Early Intentions (Dedicated to His 100th Birthday)'.
6. See Deutsch, 'Field Theory in Social Psychology', for a useful summary of Lewin's theory. Deutsch warns his readers that 'Lewin's theoretical system was never fully developed' and that, as a consequence, 'there is a vagueness and lack of precision about many concepts'. Ibid., 423.
7. Lewin, 'Frontiers in Group Dynamics', 13.
8. For a history of the Child Welfare Research Station see Cravens, *Before Head Start*.
9. See the excellent analysis of these experiments by Lezaun and Calvillo, 'In the Political Laboratory'.
10. Graebner, 'The Small Group and Democratic Social Engineering, 1900–1950'; see also Elteren, 'From Emancipating to Domesticating the Workers'.
11. Graebner, 'The Small Group and Democratic Social Engineering, 1900–1950', 150.
12. Ibid., 148.
13. Cartwright, in Patnoe, *A Narrative History of Experimental Social Psychology*, 31.
14. Zander, in ibid., 47.
15. Cartwright, in ibid., 32.
16. Bavelas, 'An Experiment in the Retraining of Recreation Leaders'.
17. Recollections of Alex Bavelas of his work with Lewin, collected by Alfred Marrow [Box M944, folder #1, Marrow papers], the Drs. Nicholas and Dorothy Cummings Center for the History of Psychology, University of Akron. Quote on p. 4.

18. Ibid., 5.
19. Ibid.
20. Bavelas and Lewin, 'Training in Democratic Leadership', 115.
21. Ibid., 116.
22. Ibid., 118.
23. Bavelas, 'Morale and the Training of Leaders'.
24. Ibid., 158.
25. Ibid., 159.
26. Ibid.
27. Ibid., 160.
28. Ibid.
29. Ibid., 161.
30. Ibid., 160.
31. Marrow, *The Practical Theorist.*
32. Bavelas, 'A Method for Investigating Individual and Group Ideology'.
33. Ibid., 374.
34. Graebner, 'The Small Group and Democratic Social Engineering, 1900–1950', 140.
35. Cartwright, 'Some Things Learned', 5.
36. Lewin, 'The Relative Effectiveness of a Lecture Method and a Method of Group Decision for Changing Food Habits', 1.
37. Ibid.
38. Ibid., 3.
39. Ibid., 7.
40. Ibid., 5.
41. Ibid., 9.
42. Ibid.
43. Ibid., 5.
44. Ibid.
45. Burnes, 'Kurt Lewin and the Harwood Studies'; Marrow, *The Practical Theorist.*
46. Lewin had done research in the textile industry before, in the textile mills in Silesia in the 1920s. John, Eckardt and Hiebsch, 'Kurt Lewin's Early Intentions (Dedicated to His 100th Birthday)'.
47. Marrow, *The Practical Theorist*, 143.
48. For a brief overview of the Harwood studies that were supervised by Lewin, see Burnes, 'Kurt Lewin and the Harwood Studies'.
49. Lewin, 'The Solution of a Chronic Conflict in a Factory'. Later republished under the slightly grander title 'The Solution of a Chronic Conflict in Industry'. Here I will use the manuscript, which is longer and contains more details of Bavelas's intervention. Lewin, 'The Solution of a Chronic Conflict in a Factory. Analysis of a Report by Alex Bavelas'.
50. This must refer to one department of the Marion plant, which in total employed some 700 people. 'Floor girls' had a function equivalent to foremen, i.e. instructing, assisting and supervising the operators. The 'supervisor', Sulinda in this case, was the head floor girl. See e.g. Oxtoby, 'An Observational Method for Determining the Job of Individual Floor Girls'.

51. Lewin, 'The Solution of a Chronic Conflict in a Factory. Analysis of a Report by Alex Bavelas', 2.
52. Ibid., 2–3.
53. Ibid., 4.
54. Ibid., 8.
55. Ibid., 13.
56. Ibid.
57. Ibid.
58. Ibid., 23.
59. Ibid., 10.
60. Ibid., 12.
61. Ibid., 15.
62. Ibid., 16.
63. Ibid., 19.
64. Ibid., 20.
65. Ibid., 25.
66. Ibid., 29.
67. Ibid., 28.
68. Ibid., 31–32.
69. Ibid., 32.
70. Billig, 'The Myth of Kurt Lewin and the Rhetoric of Collective Memory in Social Psychology Textbooks'.
71. Ibid.
72. The manuscript also includes formulas and references to figures; the latter haven't survived.
73. Lewin, 'The Solution of a Chronic Conflict in a Factory. Analysis of a Report by Alex Bavelas', 32.
74. Ibid.
75. Bavelas, 'A Method for Investigating Individual and Group Ideology'.
76. Lewin, 'The Relative Effectiveness of a Lecture Method and a Method of Group Decision for Changing Food Habits', 7.
77. Ibid.
78. Bavelas's successor at the Marion plant, John French, used role-playing to train supervisors in democratic leadership. French, 'Role-Playing as a Method of Training Foremen'.
79. Lezaun and Calvillo, 'In the Political Laboratory', 444.
80. Ibid.
81. Ibid., 443.
82. Price, 'Interview with Mason Haire'.
83. Bavelas, 'Some Mathematical Properties of Psychological Space'.
84. Collins, 'From H = Log S(n) to Conceptual Framework'; Hayles, 'Boundary Disputes'.
85. Bavelas, 'Communication Patterns in Task-Oriented Groups'.
86. Marrow, *The Practical Theorist*.
87. Capshew, *Psychologists on the March*.

88. McGregor, cited in ibid., 191. McGregor would later introduce the influential management concepts 'theory X' and 'theory Y'. McGregor, 'The Human Side of Enterprise'.
89. The AJC offered a grant of $25,000 annually over ten years for the Research Center. Miller, 'Letter from Rabbi Irving Miller'.
90. Lewin, 'The Research Center for Group Dynamics at Massachusetts Institute of Technology'.
91. Roosevelt, cited in ibid., 126. The quote is from an address by Roosevelt to the Young Democratic Club in Baltimore on April 13, 1936. When Lewin's text was reprinted two years later, after his death, editor Ronald Lippitt omitted this quote, perhaps because he thought the reference to social control sounded too authoritarian.
92. Ibid., 129.
93. Ibid., 134.
94. Ibid.
95. Ibid.
96. Cartwright, 'Achieving Change in People', 381.
97. Ibid.
98. Ibid.
99. Ibid.
100. Ibid., 382.
101. Ibid., 387.
102. Ibid., 388.
103. Ibid., 390.
104. Cartwright, 'Social Psychology in the United States during the Second World War', 348.
105. Cartwright, 'Achieving Change in People', 391.
106. The technique of role-playing, or psychodrama, had originally been developed by Jacob Moreno. Ronald Lippitt and Jack French had followed classes with him, and Moreno later became extremely upset that they had stolen his technique and used it for 'Machiavellian practices'. Moreno, cited in Highhouse, 'A History of the T-Group and Its Early Applications in Management Development', 279.
107. Cartwright, 'Some Things Learned', 18.
108. Cartwright, 'Achieving Change in People', 392.
109. Cartwright, 'Some Things Learned', 11.
110. Ibid., 10.
111. See, for instance, Marrow, *The Practical Theorist*; Back, *Beyond Words*; Highhouse 'A History of the T-Group and Its Early Applications in Management Development'; and the original report on the workshop in Lippitt, *Training in Community Relations*.
112. Back, *Beyond Words*, 11.
113. Ibid.
114. 'Specific Gripes'.
115. Barron, 'Case-Study of a Basic Skill Training Group', 6.

116. Ibid., 15. According to Back, it had been a conscious decision of all the trainers to 'create a sacred experience' in the last group meeting. Back, *Beyond Words*, 53.
117. 'Preliminary Report of the First National Training Laboratory on Group Development', 2.
118. Back, *Beyond Words*, 19.
119. Ibid., 58.
120. Cartwright, *The Research Center for Group Dynamics*.
121. Kelly, Dalton, and Francisco, *Interview of Alvin Zander*.
122. Back, *Beyond Words*.
123. Highhouse, 'A History of the T-Group and Its Early Applications in Management Development'.

8 Mind Control

1. Meerloo, *The Rape of the Mind*, 1956, 71.
2. See Carruthers, *Cold War Captives*; Young, *Name, Rank, and Serial Number*; Marks, *The Search for the 'Manchurian Candidate'*.
3. Although it is perhaps worth mentioning at this point that there is also a mind control, hypnosis and brainwashing porn genre.
4. About that entwinement, see Killen and Andriopoulos, 'Editors' Introduction on Brainwashing', and the articles in the special issue of *Grey Room* they edited. Killen, 'Homo Pavlovius', 50, calls it 'the feedback loop between paranoid fiction and paranoid practice that marked the entire Cold War'.
5. Melley, 'Brain Warfare'.
6. Scheflin and Opton, *The Mind Manipulators*.
7. Young, *Name, Rank, and Serial Number*.
8. Ibid., 187.
9. Melley, 'Brain Warfare', 31.
10. It is the subtitle of *Brain-Washing in Red China*, his first book.
11. Hunter, *Brain-Washing in Red China*.
12. Ibid., 10.
13. Ibid., 11.
14. 'Xi'nao' is the current transcription. Hunter used 'hsi nao'.
15. Lifton, '"Thought Reform" of Western Civilians in Chinese Communist Prisons', 173.
16. Ping, *The Thought Remolding Campaign of the Chinese Communist Party-State*.
17. Hunter, *Brain-Washing in Red China*, 38.
18. Ibid., 58.
19. Since then, the idea that techniques such as group decision and sensitivity training are a form of brainwashing has become a staple of conspiracy theories. See, for instance, www.crossroad.to/Quotes/brainwashing/kurt-lewin-change.htm; Griffin, *Seduction of a Generation – Sensitivity Training as Brainwashing*; and Story, *The New Underworld Order*. The resemblance between Chinese thought reform techniques and Lewinian group discussion was noted by Barlow, 'Mass Line Leadership and Thought Reform in China', and

contested (because there is no coercion in Lewin's group discussion) by Freedman and Freedman, 'Mass Line Leadership'; and McCauley, 'More about Thought Reform'. According to Lifton, *Thought Reform and the Psychology of Totalism*, 388, Lewin's work had little influence in China.

20. Hunter, *Brain-Washing in Red China*, 10.
21. In his autobiography, Meerloo seems keen to claim priority, pointing out that his article came out 'February 8, 1951' and Hunter's book followed 'a year later' Meerloo, *Een mond vol spijkers*, 205. In fact, Hunter's *Brain-Washing in Red China* came out in the same year.
22. Meerloo, *Een mond vol spijkers*.
23. Meerloo, 'The Crime of Menticide', 594.
24. Ibid., 595.
25. Ibid.
26. Ibid.
27. Ibid., 596.
28. Ibid.
29. Ibid., 597.
30. Ibid., 596.
31. Ibid., 597.
32. Ibid. For the hypnosis hype in the 1950s and the therapeutic fashion of 'hypnoanalysis', see Genter, '"Hypnotizzy" in the Cold War'.
33. Meerloo, 'The Crime of Menticide', 597.
34. Ibid.
35. See Carruthers, *Cold War Captives*, and especially Young, *Name, Rank, and Serial Number*, about prisoners of war in the Korean War. Both sides exploited their prisoners for propaganda purposes, as a result of which the ceasefire negotiations dragged on, effectively doubling the duration of the war.
36. Anthony Leviero, quoted in Carruthers, *Cold War Captives*, 201.
37. Memorandum from Dulles for the Under Secretary of State and the Deputy Secretary of Defense, subject Brainwashing, undated, Document Number (FOIA)/ESDN (CREST): CIA-RDP80B01676R002600130048-0.pdf, p. 1.
38. Melley 'Brain Warfare', 30, puts it down to the 'hypercompartmentalization' of the CIA: Dulles wasn't in on the secret.
39. Ibid., 2.
40. Ibid., 2.
41. Ibid., 3.
42. Memorandum from PSB committee, 'Planning for Psychological Exploitation of Communist Methods of Indoctrination and Inducing Confessions', for: Allen Dulles, subject: Report, undated, Document Number (FOIA)/ESDN (CREST): CIA-RDP80B01676R002600130048-0, p. 2. Secretary of Defense Charles Wilson had suggested accusing the Chinese of 'menticide' as 'effective counterpropaganda' against the germ warfare accusations. Wilson, quoted in Young, *Name, Rank, and Serial Number*, 179.
43. 'Suggested Guidance', PSB D-41, April 3, 1953, Document Number (FOIA)/ESDN (CREST): RDP80R01731R003200100001-9, p. 3.

44. Ibid., 4.
45. Brain Warfare, April 6, 1953, Document Number (FOIA)/ESDN (CREST): CIA-RDP80R01731R001700030015-9, p. 1. An excerpt was published in *U.S. News and World Report*, May 8, 1953.
46. Ibid., 1.
47. Ibid., 7.
48. Young, *Name, Rank, and Serial Number*.
49. Ibid., 113.
50. National Operations Plan to Exploit Bacteriological Warfare Hoax, Mistreatment of Prisoners of War, and Other Atrocities Perpetrated by Communist Forces during the Korean War, August 2, 1953. Document Number (FOIA)/ESDN (CREST): cia-rdp80r01731r001300220009-9, p. 3.
51. 'Basic Plan', included in the same document, p. 2.
52. 'National Plan for Exploiting Communist Mistreatment of U.N. Prisoners of War', included in the same document, p. 3.
53. Young, *Name, Rank, and Serial Number*, 129.
54. Mayo and United States Delegation to the General Assembly, Statement by the Honorable Charles W. Mayo, United States Alternate Representative on the Political Committee on 'The Question of Impartial Investigation of Charges of Use by United Nations Forces of Bacteriological Warfare', 8.
55. Ibid.
56. Ibid.
57. Ibid.
58. Ibid., 9.
59. Carruthers, *Cold War Captives*, 293 note 83.
60. Melley, 'Brainwashed! Conspiracy Theory and Ideology in the Postwar United States'; Melley, 'Brain Warfare'; Seed, *Brainwashing*.
61. *Court-Martial Reports. Holdings and Decisions of the Judge Advocates General Boards of Review and United States Court of Military Appeals*, 19:489.
62. Killen, 'Homo Pavlovius', 44, writes that 'for Hunter, understanding the true nature of this phenomenon [brainwashing] entailed placing Pavlov at its center' but fails to mention that Pavlov is absent in Hunter's first book.
63. Hunter, *Brainwashing; the Story of Men Who Defied It*, 22.
64. Ibid., 25.
65. Ibid., 243.
66. Ibid., 201.
67. Ibid., 203.
68. Ibid., 27.
69. Ibid., 172.
70. Ibid., 287.
71. Ibid., 269.
72. Ibid., 16.
73. Lutz, 'Epistemology of the Bunker', 258, writes that Meerloo at this point 'dramatised brainwashing further as "menticide"', but Meerloo had already written about menticide in 1951.
74. 'Torture Role Cited in Germ Confession', *Deseret News*, March 10, 1954.
75. Meerloo, 'Pavlovian Strategy as a Weapon of Menticide', 809.

76. Ibid., 810.
77. Ibid., 811.
78. Ibid., 812.
79. Meerloo, *The Rape of the Mind*, 1956, 35.
80. Ibid., 212.
81. Ibid., 211.
82. Ibid., 212.
83. Killen, 'Homo Pavlovius', explores the association between Pavlov, brainwashing and television in the 1950s.
84. Meerloo, *The Rape of the Mind*, 1956, 54.
85. Ibid., 90.
86. Ibid., 14.
87. Ibid., 102.
88. Ibid., 211.
89. Ibid., 294.
90. Ibid., 296.
91. Ibid., 97.
92. Dollard, 'Men Who Are Tortured by the Awful Fear of Torture'.
93. Schein, 'Knuckling Under', 140.
94. Schein worked at the Walter Reed Army Institute of Research, see Lutz, 'Epistemology of the Bunker: The Brainwashed and Other New Subjects of Permanent War'.
95. Hinkle and Wolff, 'Communist Control Techniques', 113.
96. Schein, 'The Chinese Indoctrination Program for Prisoners of War', 172.
97. Bauer, 'Brainwashing', 41.
98. Biderman, 'Communist Attempts to Elicit False Confessions from Air Force Prisoners of War', 617.
99. Hinkle and Wolff, 'Brainwashing from a Psychological Viewpoint', v.
100. Ibid., i.
101. Ibid.
102. Bauer and Schein, 'Introduction', 4.
103. Hinkle and Wolff, 'Brainwashing from a Psychological Viewpoint'.
104. Bauer worked at Harvard's Russian Research Project (funded by Carnegie Corporation, Air Force and CIA) and MIT's Center for International Studies (CENIS), which was also funded a.o. by the CIA; see Lutz, 'Epistemology of the Bunker'.
105. Bauer, 'Brainwashing', 41.
106. Hinkle and Wolff, 'The Methods of Interrogation and Indoctrination Used by the Communist State Police', 610.
107. Hinkle and Wolff, 'Brainwashing from a Psychological Viewpoint', 66–67.
108. Lifton, 'Chinese Communist "Thought Reform"', 639.
109. Hinkle and Wolff described the Russian procedure in greatest detail – Hinkle and Wolff, 'Brainwashing from a Psychological Viewpoint'; Hinkle and Wolff, 'Communist Control Techniques'; Hinkle and Wolff, 'Communist Interrogation and Indoctrination of "Enemies of the State"'; Hinkle and Wolff, 'The Methods of Interrogation and Indoctrination Used by the Communist State Police'.

110. About the Chinese re-education procedure, see Lifton, '"Thought Reform" of Western Civilians in Chinese Communist Prisons'; Lifton, 'Chinese Communist "Thought Reform"'. Schein, 'The Chinese Indoctrination Program for Prisoners of War'; and Biderman, 'Communist Attempts to Elicit False Confessions from Air Force Prisoners of War', describe how the Chinese operated in Korea.
111. Lifton, 'Chinese Communist "Thought Reform"', 631.
112. Biderman, 'Communist Attempts to Elicit False Confessions from Air Force Prisoners of War'.
113. Hinkle and Wolff, 'Brainwashing from a Psychological Viewpoint', 79.
114. Lifton, '"Thought Reform" of Western Civilians in Chinese Communist Prisons', 186.
115. Lifton, 'Chinese Communist "Thought Reform"', 633.
116. Ibid., 640.
117. Schein, Schneier and Barker, *Coercive Persuasion.*
118. The first two were versions of their internal CIA report; the third and fourth summarised the findings for an academic audience. Their report was widely considered as the canonical account of brainwashing, see Marks, *The Search for the 'Manchurian Candidate'*.
119. Hinkle and Wolff, 'Brainwashing from a Psychological Viewpoint', 6.
120. Ibid., 75.
121. Ibid., 66.
122. Hinkle and Wolff, 'Communist Control Techniques', 155.
123. Hinkle and Wolff, 'Communist Interrogation and Indoctrination of "Enemies of the State"', 116.
124. Ibid., 169.
125. Hinkle, 'Communist Manipulation of Behavior', 1913.
126. Hinkle and Wolff, 'The Methods of Interrogation and Indoctrination Used by the Communist State Police', 607.
127. Schein, 'The Chinese Indoctrination Program for Prisoners of War', 158.
128. Hinkle and Wolff, 'Brainwashing from a Psychological Viewpoint', 63.
129. Lifton, 'Chinese Communist "Thought Reform"', 643.
130. Ibid.
131. Ibid.
132. Ibid., 644.
133. Ibid., 638.
134. Ibid., 642.
135. A conclusion that Hinkle and Wolff shared: 'So called "brain washing" produces no permanent changes in the function of the brain.' 'Communist Control Techniques', 109.
136. Schein, Schneier and Barker, *Coercive Persuasion.*
137. Lemov's claim that the experts were 'near-unanimous' in concluding that brainwashing was done by a combination of environmental control and 'intensive behavioral conditioning' is wrong. Lemov, 'Brainwashing's Avatar', 71. On the contrary: they were near-unanimous in denying a role for conditioning. Equally wrong is Melley's claim that Biderman, Schein and Lifton concluded that brainwashing was nothing more than 'a

brutal combination of isolation, physical deprivation, and the nearly interminable revision of a personal confession'. Melley, 'Brain Warfare', 29.

138. Melley, 'Brainwashed! Conspiracy Theory and Ideology in the Postwar United States', 149.
139. Hunter, *Brainwashing; the Story of Men Who Defied It*, 255.
140. Hinkle and Wolff, 'Communist Interrogation and Indoctrination of "Enemies of the State"', 123. The same was true of the Chinese system: it maintained that 'the Government never arrests innocent people'. Lifton, 'Chinese Communist "Thought Reform"', 638.
141. Hinkle and Wolff, 'The Methods of Interrogation and Indoctrination Used by the Communist State Police', 139.
142. Hinkle and Wolff, 'Communist Interrogation and Indoctrination of "Enemies of the State"'.
143. Lifton, 'Chinese Communist "Thought Reform"', 643.
144. Lifton, '"Thought Reform" of Western Civilians in Chinese Communist Prisons', 189, quoting Florence Kluckhohn and John Spiegel. Lifton also referred to the work of Gregory Bateson.
145. Miller, 'Brainwashing', also described brainwashing in systems terms.
146. Hinkle and Wolff, 'Communist Interrogation and Indoctrination of "Enemies of the State"', 174.
147. Pavlovian principles, for example. Killen, 'Homo Pavlovius', notes that Wolff had stayed at Pavlov's laboratory in the 1930s and was the founder of the Pavlovian Society. In his own research projects for the CIA he applied Pavlovian theories of conditioning. Hinkle and Wolff, 'Brainwashing from a Psychological Viewpoint', 63.
148. Marks, *The Search for the 'Manchurian Candidate'*, 54–55.
149. Ibid.
150. Ibid., 28.
151. Janis, 'Are the Cominform Countries Using Hypnotic Techniques to Elicit Confessions in Public Trials?'
152. The American hypnotist George Estabrooks was an important source of spectacular claims about the powers of hypnosis.
153. See Andriopoulos, 'The Sleeper Effect', about the 'sleeper effect'; Andriopoulos connects remote control by post-hypnotic suggestion with the sleeper effect in persuasion as studied by Hovland and with undercover 'sleeper agents'.
154. Janis, 'Are the Cominform Countries Using Hypnotic Techniques to Elicit Confessions in Public Trials?', 19.
155. Ibid., 20.
156. John Marks's *The Search for the 'Manchurian Candidate'* remains one of the best studies of the CIA mind control projects. He estimates that MKULTRA cost the US taxpayers about $10 million. Marks, *The Search for the 'Manchurian Candidate'*, 197. See also McCoy, *A Question of Torture*; and Collins, *In the Sleep Room*.
157. Cited in Marks, *The Search for the 'Manchurian Candidate'*, 49.

158. Memorandum to [deleted], from [deleted], subject Special Research, BLUEBIRD, January 2, 1952. Document Number (FOIA)/ESDN (CREST): 0000140401.
159. Ibid., 6.
160. Document dated February 25, 1952, Mr. R. L. Bannerman, Mr. Morse Allen, report of interview with anonymised hypnotist, February 21, 1952, p. 1. Document code A/B/,III, 6, 7 ARTICHOKECIAfiles.pdf, www.paperlessarchives.com.
161. Ibid., 7.
162. Ibid.
163. Ibid., 5.
164. Ibid., 8.
165. Memorandum from Morse Allen via R.L. Bannerman to [deleted], dated March 11, 1952, reporting an interview of March 7, 1952, with anonymised person. ARTICHOKECIAfiles.pdf, www.paperlessarchives.com.
166. Ibid., 2
167. Marks, *The Search for the 'Manchurian Candidate'*, 162.
168. About Cameron see Marks, *The Search for the 'Manchurian Candidate'*; McCoy, *A Question of Torture*; Lemov, 'Brainwashing's Avatar'; and particularly Collins, *In the Sleep Room*.
169. Collins quotes a former assistant of Cameron who claimed he got the idea for psychic driving from Huxley's *Brave New World*. Collins, *In the Sleep Room*, 123.
170. Cameron, 'Psychic Driving', 508.
171. Cameron, 'Adventures with Repetition', 315.
172. Cameron, 'Psychic Driving', 503.
173. Ibid., 508.
174. Ibid.
175. Ibid.
176. 'Private Institutions Used in C.I.A. Effort to Control Behavior', 16.
177. Quoted in Collins, *In the Sleep Room*, 135.
178. The development of psychic driving continued, however; see Cameron et al., 'Automation of Psychotherapy', for its final form.
179. Collins, *In the Sleep Room*.
180. Cameron and Pande, 'Treatment of the Chronic Paranoid Schizophrenic Patient', 93.
181. Cameron, 'Production of Differential Amnesia as a Factor in the Treatment of Schizophrenia'.
182. Ibid., 28.
183. Collins, *In the Sleep Room*.
184. See ibid. In the mid-1960s CIA psychiatrists conducted further electroshock experiments using Vietcong prisoners. McCoy, *A Question of Torture*. These too resulted in failure.
185. Cameron and Pande, 'Treatment of the Chronic Paranoid Schizophrenic Patient', 93.

186. Cameron, Lohrenz and Handcock, 'The Depatterning Treatment of Schizophrenia', 69.
187. Cameron, 'Adventures with Repetition', 321.
188. Marks, *The Search for the 'Manchurian Candidate'*, 152.
189. 'KUBARK' was the CIA's cryptonym for itself. Kleinman, 'KUBARK Counterintelligence Interrogation Review', 96, claims it stood for a specific CIA operation in the early 1960s, but he appears to be alone in this view.
190. McCoy, *A Question of Torture*, 7.
191. See Kleinman, 'KUBARK Counterintelligence Interrogation Review', for a reading of the KUBARK manual that is similar to mine.
192. CIA, 'KUBARK Counterintelligence Interrogation', 111.
193. Jeffrey Kaye, of the investigative news site MuckRock, requested the report under the Freedom of Information Act, but the Pentagon eventually wrote back that a comprehensive search had uncovered no such record; see www.muckrock.com/foi/united-states-of-america-10/request-for-af-study-on-developing-improved-interrogation-techniques-8913/.
194. Published two years later in the journal *Sociometry*, Biderman, 'Social-Psychological Needs and "Involuntary" Behavior as Illustrated by Compliance in Interrogation'.
195. Ibid., 125.
196. Ibid., 143.
197. CIA, 'KUBARK Counterintelligence Interrogation', 12.
198. Ibid., 10.
199. Ibid., 48.
200. Ibid., 20.
201. Ibid., 42.
202. Ibid., 12.
203. Ibid., 41.
204. Ibid., 92.
205. Ibid., 51.
206. Meerloo, *The Rape of the Mind*, 1956, 71.
207. Morawski, 'Epistemological Dizziness in the Psychology Laboratory'.
208. Ibid., 596.
209. Ian Nicholson has argued that Milgram knowingly 'tormented' the subjects in his obedience experiments. Nicholson, 'Torture at Yale'. I agree, but that study is an exception. Nicholson also describes the experimental situation in Milgram's study as one in which the experimenter had 'total control'. Ibid., 753, but that seems to me a mischaracterisation, given that the manipulation was not always successful.
210. Moreno, *Mind Wars*.
211. McCoy, *A Question of Torture*.
212. Eakin and Danner, 'Our New Politics of Torture'; Konnikova, 'Trying to Cure Depression, but Inspiring Torture'; and Shaw, 'The Psychologists Take Power'; among others, have reported on this affair.
213. Senate Select Committee on Intelligence, 'The Senate Intelligence Committee Report on Torture'.

214. For what it's worth, Bruce Mitchell has contested the conclusions of the Senate committee, claiming among other things that the interrogations involved rapport building if prisoners were cooperative and that waterboarding was effective in making the resisters talk. Leopold, 'Psychologist James Mitchell Admits He Waterboarded Al Qaeda Suspects'.

9 The Priming Saga: The Subtle Technology of Psychological Experimentation

1. Morawski, 'Epistemological Dizziness in the Psychology Laboratory', 575.
2. The discussion about Jonah Lehrer's *New Yorker* piece 'The Truth Wears Off' in 2010 was an earlier indication of the unease among psychologists about the state of their discipline. Lehrer described the unsettling phenomenon that a number of experimental effects in psychology and other sciences seemed to get smaller in subsequent replication studies. Lehrer, 'The Truth Wears Off'.
3. Bem, 'Feeling the Future'.
4. A journal's impact factor is a measure of how often the articles that it publishes are cited in other articles. The *JPSP* has an impact factor of 5.031. The number-one journal in social psychology is the *Personality and Social Psychology Review*, boasting an IF of 6.692. (Both impact factors as per March 30, 2016.)
5. Bem, 'Feeling the Future', 409.
6. Wagenmakers et al., 'Why Psychologists Must Change the Way They Analyze Their Data', 426.
7. Judd and Gawronski, 'Editorial Comment', 406.
8. Ibid.
9. Bem, 'Feeling the Future', 420.
10. Ritchie, Wiseman and French, 'Replication, Replication, Replication'.
11. Ritchie, Wiseman and French, 'Failing the Future'.
12. Levelt, Groenhuijsen and Hagenaars, 'Interim-rpportage inzake door Prof. Dr. D.A. Stapel gemaakte inbreuk op wetenschappelijke integriteit', 13.
13. Simonsohn, 'Just Post It'.
14. Simmons, Nelson and Simonsohn, 'False-Positive Psychology'.
15. Ibid., 1359.
16. John, Loewenstein and Prelec, 'Measuring the Prevalence of Questionable Research Practices with Incentives for Truth-Telling'.
17. 0.6 per cent of respondents admitted to having falsified data. However, 22 per cent reported having rounded down the all-important p-value – 0.054 to 0.05, for example – to make it seem the results were statistically significant at the conventional 0.05 level. Strictly speaking, this is falsification and thus fraud.
18. Bargh, 'Nothing in Their Heads'.
19. Yong, 'A Failed Replication Draws a Scathing Personal Attack from a Psychology Professor'.

20. It can still be found on the Internet Archive at http://web.archive.org/web/20120307100648/http://www.psychologytoday.com/blog/the-natural-unconscious/201203/nothing-in-their-heads.
21. Nosek, Spies and Motyl, 'Scientific Utopia', 616.
22. Ferguson and Heene, 'A Vast Graveyard of Undead Theories', 555.
23. Open Science Collaboration, 'An Open, Large-Scale, Collaborative Effort to Estimate the Reproducibility of Psychological Science'.
24. Ibid., 657.
25. Nosek and Bar-Anan, 'Scientific Utopia'; Nosek, Spies and Motyl, 'Scientific Utopia'.
26. Fiedler, Kutzner and Krueger, 'The Long Way from α-Error Control to Validity Proper', 661.
27. Stroebe, Postmes and Spears, 'Scientific Misconduct and the Myth of Self-Correction in Science', 679.
28. Bargh, Chen and Burrows, 'Automaticity of Social Behavior', 236.
29. Ibid., 237.
30. Ibid., 236.
31. Ibid., 230.
32. Ibid., 231.
33. See Chapter 5.
34. Nathaniel Rivers and I discuss priming research in our exploration of deception in rhetoric: Rivers and Derksen, 'Ecologies of Deception'.
35. Ironically, Bem too put his precognition experiments in the context of the study of automatic, unconscious processes in social psychology; one of his experiments concerned 'retroactive priming'. Bem, 'Feeling the Future'.
36. Bargh, Chen and Burrows, 'Automaticity of Social Behavior', 230.
37. Bargh, 'Free Will Is Un-Natural'.
38. About the rhetorical function of theories of human irrationality in the service of selling psychological expertise, see also Lopes, 'The Rhetoric of Irrationality'.
39. Karremans, Stroebe and Claus, 'Beyond Vicary's Fantasies'.
40. Pratkanis, 'The Cargo-Cult Science of Subliminal Persuasion – CSI'.
41. Stroebe, 'The Subtle Power of Hidden Messages'.
42. Doyen et al., 'Behavioral Priming', 1.
43. Ibid., 3.
44. Ibid., 6.
45. Ibid.
46. Bargh, 'Nothing in Their Heads'.
47. Ibid.
48. Bargh and Chartrand, 'The Mind in the Middle'.
49. Ibid., 259.
50. Ibid., 263.
51. Ibid., 285.
52. For example, Pashler et al., 'Priming of Social Distance?'; and Harris et al., 'Two Failures to Replicate High-Performance-Goal Priming Effects'; reported failures to replicate other priming studies of Bargh, and Shanks

et al., 'Priming Intelligent Behavior', were unable to reproduce Dijksterhuis's famous 'Trivial Pursuit' effect, mentioned in Chapter 1.

53. Yong, 'Replication Studies: Bad Copy', 299.
54. Dijksterhuis, 'Welcome Back Theory!', 73.
55. He took it off later.
56. Mitchell, 'On the Emptiness of Failed Replications'.
57. Schnall, 'Simone Schnall on Her Experience with a Registered Replication Project'.
58. Schnall, 'Moral Intuitions, Replication, and the Scientific Study of Human Nature'; Schnall, 'Simone Schnall on Her Experience with a Registered Replication Project'; Schnall, 'Social Media and the Crowd-Sourcing of Social Psychology'.
59. Lieberman, 'Latitudes of Acceptance'.
60. Kahneman, cited in Yong, 'Replication Studies: Bad Copy'.
61. Neuroskeptic, 'The Replication Crisis'.
62. Wilson, 'Psychology's Real Replication Problem'.
63. Quoted in Yong, 'Replication Studies: Bad Copy', 299.
64. Bargh, 'Our Unconscious Mind', 34.
65. Cesario, Plaks and Higgins, 'Automatic Social Behavior as Motivated Preparation to Interact.'
66. Dijksterhuis, 'Welcome Back Theory!', 72.
67. Ibid., 73–74.
68. Ibid., 74.
69. Stroebe and Strack, 'The Alleged Crisis and the Illusion of Exact Replication', 65.
70. Ibid., 62.
71. Ibid., 65.
72. Cesario, 'Priming, Replication, and the Hardest Science', 45.
73. Ibid., 43.
74. Ibid., 42.
75. Ibid., 44.
76. Ibid.
77. Ibid., 45.
78. Schröder and Thagard, 'The Affective Meanings of Automatic Social Behaviors'.
79. Ibid., 256.
80. Ibid., 257.
81. Klatzky and Creswell, 'An Intersensory Interaction Account of Priming Effects – and Their Absence', 50.
82. Ibid., 53.
83. Ibid., 54.
84. Ibid., 55.
85. Ibid., 56.
86. Ibid.
87. Locke, 'Theory Building, Replication, and Behavioral Priming', 412.
88. Schwartz, 'On Klatzky and Creswell (2014): Saving Social Priming Effects but Losing Science as We Know It?', 405.

89. Ibid., 406.
90. Ibid., 405.
91. Fitzsimons, Chartrand and Fitzsimons, 'Automatic Effects of Brand Exposure on Motivated Behavior'.
92. Srivastava, 'An Open Review of Many Labs', 3.
93. One of the projects has now been published; see Klein et al., 'Investigating Variation in Replicability'.
94. Richard Kunert has argued that the unknown moderator account implies that studies that have been conceptually replicated should also have more success in direct replications because their results have already proven to be somewhat robust. Analysis of the data of the Reproducibility Project showed that there is no correlation and that QRPs are therefore a more likely explanation of nonreplications. Kunert, 'Internal Conceptual Replications Do Not Increase Independent Replication Success'.
95. Gelman and Loken, 'The Statistical Crisis in Science'.
96. Ibid., 461.
97. Gelman, 'Primed to Lose'.
98. Radder, 'Science, Technology and the Science–Technology Relationship', 81.
99. Ibid., 82.
100. Stroebe and Strack acknowledge the importance of direct replication in applied research: 'A scientist who wants to establish the efficiency of a specific treatment or intervention is well advised to repeatedly apply exactly the same procedure.' Stroebe and Strack, 'The Alleged Crisis and the Illusion of Exact Replication', 60.
101. Billig, 'Rhetoric of Social Psychology', 59.
102. Radder, 'Causality, Theory Ladenness and Reproducibility in Experimental Science'; Radder, 'De cognitieve drie-eenheid'.
103. Radder, 'Causality, Theory Ladenness and Reproducibility in Experimental Science', 8.
104. Ruud Abma argues along similar lines that the experimental method is of limited use in psychology, but he does not reject the experiment completely. Abma, *Over de grenzen van disciplines*; Abma, 'Replicaties in de psychologie – reactie op Radder'.
105. Radder, 'Causality, Theory Ladenness and Reproducibility in Experimental Science', 8.
106. Ibid., 9.
107. Gergen, 'Social Psychology as History'.
108. Ibid., 310.
109. Srivastava, 'Making Progress in the Hardest Science'.
110. Schlenker, 'Social Psychology and Science'.
111. For example, in their defense of experimental social psychology, Stroebe and Strack refer to Gergen's article to underscore their point that direct replication is not feasible in basic social psychological research and then present conceptual replication as the strategy of choice. Stroebe and Strack, 'The Alleged Crisis and the Illusion of Exact Replication'.

112. Derksen and Rietzschel, 'Surveillance Is Not the Answer, and Replication Is Not a Test'.
113. In a similar vein, Steve Brown and Paul Stenner, discussing the work of Michel Serres, have written that psychology has 'a preference for the preformed and the ever-repeated: a preference for the universal that is discovered via experiments that can be perfectly replicated'. Instead, they wish for psychology to attend to 'production, composition and combination'. I would argue that psychology has been seeking the universal via experiments that were not supposed to be perfectly replicable and that it is precisely the current move towards direct replication that is productive. Brown and Stenner, *Psychology without Foundations*, 44.

10 Conclusion

1. Owen, *A New View of Society*, 34.
2. Hunter, *Brain-Washing in Red China*.
3. Sociologist of science Harry Collins would argue that the kind of expertise that is required for tact is always in part tacit and requires being socialised in the culture in question. His latest exposition of this 'social cartesianism' is Collins, *Tacit and Explicit Knowledge*.
4. 'NeuroStandards Project White Paper'.
5. Ibid., 32.

Bibliography

Abbott, Andrew. 'Pragmatic Sociology and the Public Sphere: The Case of Charles Richmond Henderson'. *Social Science History* 34, no. 3 (September 21, 2010): 337–71.

Abma, Ruud. *Over de grenzen van disciplines: plaatsbepaling van de sociale wetenschappen*. Nijmegen: Vantilt, 2011.

'Replicaties in de psychologie – reactie op Radder'. *Algemeen Nederlands Tijdschrift Voor Wijsbegeerte* 106, no. 1 (2014): 55–61.

Akin, William E. *Technocracy and the American Dream: The Technocrat Movement, 1900–1941*. Oakland: University of California Press, 1977.

Alexander, Jack. 'The Green Pencil'. *The New Yorker*, December 11, 1937.

Andriopoulos, Stefan. 'The Sleeper Effect: Hypnotism, Mind Control, Terrorism'. *Grey Room*, no. 45 (Fall 2011): 88–105.

Anonymous. 'The Mistakes of the Efficiency Men'. In *Scientific Management. A Collection of the More Significant Articles Describing the Taylor System of Management*, edited by C. B. Thompson, 615–31. Easton: Hive Pub. Co., 1972.

Back, Kurt Wolfgang. *Beyond Words: The Story of Sensitivity Training and the Encounter Movement*. New York: Russell Sage Foundation, 1972.

Baistow, K. 'Behavioural Approaches and the Cultivation of Competence'. In *Psychology in Britain. Historical Essays and Personal Reflections*, edited by G. C. Bunn, A. D. Lovie and G. D. Richards, 309–29. London: Science Museum, British Psychological Society, 2001.

Baker, K. M. 'The Early History of the Term "Social Science"'. *Annals of Science* 20, no. 3 (1964): 211–260.

Bannister, Robert C. *Sociology and Scientism: The American Quest for Objectivity, 1880–1940*. Chapel Hill: University of North Carolina Press, 1987.

Bargh, J. A. 'Free Will Is Un-Natural'. In *Are We Free?: Psychology and Free Will*, edited by John Baer, J. C. Kaufman and R. E. Baumeister, 128–54. New York: Oxford University Press, 2008.

'Nothing in Their Heads'. *The Natural Unconscious*, March 5, 2012. http://web.archive.org/web/20120307100648/http://www.psychologytoday.com/blog/the-natural-unconscious/201203/nothing-in-their-heads.

'Our Unconscious Mind'. *Scientific American* 310, no. 1 (2014): 30–37.

Bargh, J. A., and T. L. Chartrand. 'The Mind in the Middle: A Practical Guide to Priming and Automaticity Research'. In *Handbook of Research Methods in Social and Personality Psychology*, edited by H. T. Reis and C. M. Judd, 253–87. Cambridge: Cambridge University Press, 2000.

Bargh, John A., Mark Chen and Lara Burrows. 'Automaticity of Social Behavior: Direct Effects of Trait Construct and Stereotype Activation on Action'. *Journal of Personality and Social Psychology* 71, no. 2 (1996): 230–44.

Barlow, John A. 'Mass Line Leadership and Thought Reform in China'. *American Psychologist* 36, no. 3 (1981): 300–9.

Barron, Margaret. 'Case-Study of a Basic Skill Training Group', August 15, 1947. Box 220, folder #1. The Drs. Nicholas and Dorothy Cummings Center for the History of Psychology.

Bauer, Raymond A. 'Brainwashing: Psychology or Demonology?' *Journal of Social Issues* 13, no. 3 (July 1957): 41–47.

Bauer, Raymond A., and Edgar H. Schein. 'Introduction'. *Journal of Social Issues* 13, no. 3 (July 1, 1957): 3–4.

Bavelas, Alex. 'A Method for Investigating Individual and Group Ideology'. *Sociometry* 5, no. 4 (November 1, 1942): 371–77.

'An Experiment in the Retraining of Recreation Leaders'. Iowa City: University of Iowa, 1942.

'Communication Patterns in Task-Oriented Groups'. *The Journal of the Acoustical Society of America* 22, no. 6 (1950): 725–30.

'Morale and the Training of Leaders'. In *Civilian Morale: Second Yearbook of the Society for the Psychological Study of Social Issues*, edited by Goodwin Watson and Goodwin (Ed) Watson, 143–65. Boston: Houghton Mifflin Company, 1942.

'Some Mathematical Properties of Psychological Space'. Thesis, Massachusetts Institute of Technology, 1948. http://dspace.mit.edu/handle/1721.1/34690.

Bavelas, Alex, and Kurt Lewin. 'Training in Democratic Leadership'. *Journal of Abnormal and Social Psychology* 37, no. 1 (1942): 115–19.

Bem, D. J. 'Feeling the Future: Experimental Evidence for Anomalous Retroactive Influences on Cognition and Affect'. *Journal of Personality and Social Psychology* 100, no. 3 (2011): 407–25.

Benjafield, John G. *A History of Psychology*. Don Mills, Ontario; New York: Oxford University Press, 2005.

Benjamin, Ludy T., and David B. Baker. *From Séance to Science: A History of the Profession of Psychology in America*. Australia; Belmont, CA: Thomson/Wadsworth, 2004.

Bernard, L. L. 'Some Dynamics of Present World Relations'. *Phylon (1940–1956)* 7, no. 2 (June 1, 1946): 105–19.

'Standards of Living and Planes of Living'. *Social Forces* 7, no. 2 (December 1, 1928): 190–202.

'The Place of the Social Sciences in Modern Education'. *Journal of Educational Sociology* 9, no. 1 (1935): 47–55.

'The Teaching of Sociology in the United States'. *American Journal of Sociology* 15, no. 2 (October 1, 1909): 164–213.

Biderman, A. 'Communist Attempts to Elicit False Confessions from Air Force Prisoners of War'. *Bulletin of the New York Academy of Medicine* 33, no. 9 (1957): 616–25.

'Social-Psychological Needs and "Involuntary" Behavior as Illustrated by Compliance in Interrogation'. *Sociometry* 23, no. 2 (June 1, 1960): 120–47.

Billig, M. *Arguing and Thinking: A Rhetorical Approach to Social Psychology*. European Monographs in Social Psychology. Cambridge: Cambridge University Press, 1987.

'Rhetoric of Social Psychology'. In *Deconstructing Social Psychology*, edited by J. Shotter and I. Parker, 47–60. London: Routledge, 1990.

'The Myth of Kurt Lewin and the Rhetoric of Collective Memory in Social Psychology Textbooks'. *Theory & Psychology*, 25, no. 6 (2015).

Blackman, Frank W. 'Reasonable Department of Sociology for Colleges and Universities'. *American Journal of Sociology* 20, no. 2 (September 1, 1914): 261–63. doi:10.2307/2763114.

Bloemen, Erik Sybrand Arnold. *Scientific Management in Nederland, 1900–1930*. Amsterdam: NEHA, 1988.

Boese, Franz. *Geschichte des Vereins für Sozialpolitik: 1872–1932*. Berlin: Duncker & Humblot, 1939.

Bogardus, Emory Stephen. *A History of Social Thought*. Los Angeles: University of Southern California Press, 1922.

Introduction to Sociology. 2nd rev. ed. Los Angeles: University of Southern California Press, 1917.

Breslau, Daniel. 'The Scientific Appropriation of Social Research'. *Theory and Society* 19, no. 4 (August 1, 1990): 417–46.

Brown, JoAnne. *The Definition of a Profession: The Authority of Metaphor in the History of Intelligence Testing, 1890–1930*. Princeton, NJ: Princeton University Press, 1992.

Brown, Steve D., and Paul Stenner. *Psychology without Foundations: History, Philosophy and Psychosocial Theory*. 1st ed. Thousand Oaks, CA: SAGE Publications Ltd., 2009.

Bryson, Dennis. 'Lawrence K. Frank, Knowledge, and the Production of the "Social"'. *Poetics Today* 19, no. 3 (October 1, 1998): 401–21.

Bulmer, Martin. *The Chicago School of Sociology: Institutionalization, Diversity, and the Rise of Sociological Research*. Chicago: University of Chicago Press, 1986.

Bulmer, Martin, and Joan Bulmer. 'Philanthropy and Social Science in the 1920s: Beardsley Ruml and the Laura Spelman Rockefeller Memorial, 1922–29'. *Minerva* 19, no. 3 (1981): 347–407.

Bunge, Mario. 'The Seven Pillars of Popper's Social Philosophy'. *Philosophy of the Social Sciences* 26, no. 4 (December 1, 1996): 528–56.

Bunn, G. C. *The Truth Machine: A Social History of the Lie Detector*. Baltimore: Johns Hopkins University Press, 2012.

Burgess, E. W., and Robert Ezra Park. *Introduction to the Science of Sociology*. Chicago: University of Chicago Press, 1921.

Burnes, Bernard. 'Kurt Lewin and the Harwood Studies: The Foundations of OD'. *The Journal of Applied Behavioral Science* 43, no. 2 (June 1, 2007): 213–31.

Cain, Susan. *Quiet: The Power of Introverts in a World That Can't Stop Talking*. London: Penguin Books, 2013.

Callon, Michel, and John Law. 'On Qualculation, Agency, and Otherness'. *Environment and Planning D: Society and Space* 23, no. 5 (October 1, 2005): 717–33.

Callon, Michel, and Fabian Muniesa, 'Economic Markets as Calculative Collective Devices'. *Organization Studies* 26, no. 8 (2005): 1229–50.

Cameron, D. E. 'Adventures with Repetition: The Search for Its Possibilities'. *Proceedings of the Annual Meeting of the American Psychopathological Association* 53 (1965): 312–22.

'Production of Differential Amnesia as a Factor in the Treatment of Schizophrenia'. *Comprehensive Psychiatry* 1, no. 1 (February 1960): 26–34.

'Psychic Driving'. *American Journal of Psychiatry* 112, no. 7 (1956): 502–9.

Cameron, D. E., Leonard Levy, Thomas Ban and Leonard Rubenstein. 'Automation of Psychotherapy'. *YCOMP Comprehensive Psychiatry* 5, no. 1 (1964): 1–14.

Cameron, D. Ewen, John G. Lohrenz and K. A. Handcock. 'The Depatterning Treatment of Schizophrenia'. *YCOMP Comprehensive Psychiatry* 3, no. 2 (1962): 65–76.

Cameron, D. E., and S. K. Pande. 'Treatment of the Chronic Paranoid Schizophrenic Patient'. *Canadian Medical Association Journal* 78, no. 2 (1958): 92–96.

Capshew, James H. *Psychologists on the March: Science, Practice, and Professional Identity in America, 1929–1969*. Cambridge Studies in the History of Psychology. Cambridge: Cambridge University Press, 1999.

Carnagey, Dale. *Public Speaking: Instructor's Manual*. New York: Association Press, 1920.

Carnegie, Dale. *Comment se faire des amis pour réussir dans la vie*. Translated by Denise Geneix. [Paris]: Hachette, 1946.

How to Win Friends and Influence People. New York: Simon & Schuster, 1936.

Public Speaking; a Practical Course for Business Men. New York: Association Press, 1926.

Carruthers, Susan L. *Cold War Captives: Imprisonment, Escape, and Brainwashing*. Berkeley: University of California Press, 2009.

Cartwright, Dorwin. 'Achieving Change in People: Some Applications of Group Dynamics Theory'. *Human Relations* 4, no. 4 (1951): 381–92.

'Social Psychology in the United States during the Second World War'. *Human Relations* 1, no. 3 (June 1, 1948): 333–52.

'Some Things Learned: An Evaluative History of the Research Center for Group Dynamics'. *Journal of Social Issues* 14, no. S12 (December 1, 1958): 3–19.

The Research Center for Group Dynamics; a Report of Five Years' Activities and a View of Future Needs. Ann Arbor: Institute for Social Research, University of Michigan, 1950.

Cawley, R. McGreggor, and William Chaloupka. 'American Governmentality. Michel Foucault and Public Administration'. *American Behavioral Scientist* 41, no. 1 (September 1, 1997): 28–42.

Cesario, Joseph. 'Priming, Replication, and the Hardest Science'. *Perspectives on Psychological Science* 9, no. 1 (January 1, 2014): 40–48.

Cesario, J., J. E. Plaks and E. T. Higgins. 'Automatic Social Behavior as Motivated Preparation to Interact'. *Journal of Personality and Social Psychology* 90, no. 6 (2006): 893–910.

CIA. 'KUBARK Counterintelligence Interrogation'. CIA, July 1963. National Security Archive. http://nsarchive.gwu.edu/NSAEBB/NSAEBB122/index.htm#hre.

Cicero. *Brutus, Orator; Volume V.* Translated by G. L. Hendrickson and H. M. Hubbell. Cambridge: Harvard University Press, 1939.

Collins, A. 'From H = Log S(n) to Conceptual Framework: A Short History of Information'. *History of Psychology* 10, no. 1 (2007): 44–72.

In the Sleep Room: The Story of the CIA Brainwashing Experiments in Canada. Toronto, Canada: Lester & Orpen Dennys, 1988.

Collins, H. M. *Tacit and Explicit Knowledge.* Chicago: University of Chicago Press, 2010.

Condorcet, Jean-Antoine-Nicolas de Caritat, Marquis de. *Outline of an Historical View of the Progress of the Human Mind.* Dublin: John Chambers, 1796.

Cooke, Morris Llewellyn, Samuel Gompers and Fred J. Miller. *Labor, Management and Production.* Philadelphia: American Acad. of Political and Social Science, 1920.

Coon, D. J. 'Testing the Limits of Sense and Science. American Experimental Psychologists Combat Spiritualism, 1880–1920'. *American Psychologist* 47, no. 2 (1992): 143–51.

Court-Martial Reports. *Holdings and Decisions of the Judge Advocates General Boards of Review and United States Court of Military Appeals.* Vol. 19. The Judge Advocates General of the Armed Forces & The United States Court of Military Appeals, 1955.

Cravens, Hamilton. *Before Head Start: The Iowa Station & America's Children.* Chapel Hill: University of North Carolina Press, 1993.

'D&B – wetenschappelijk inzicht & gedragsbeïnvloeding in de praktijk'. Accessed July 3, 2013. www.db-abs.com/home.

De Certeau, Michel. *The Practice of Everyday Life.* Berkeley: University of California Press, 1984.

De Munnik, Martin. *De koopknop: het geheim van het consumentenbrein: inclusief meer dan 50 neuromarketingtips.* Edited by Annet Talsma. Den Haag: Academic Service, 2012.

Dent, Eric B., and Susan Galloway Goldberg. 'Challenging "Resistance to Change"'. *Journal of Applied Behavioral Science* 35, no. 1 (1999): 25–41.

Derksen, M. 'Are We Not Experimenting Then? The Rhetorical Demarcation of Psychology and Common Sense'. *Theory & Psychology* 7, no. 4 (August 1997): 435–56.

'Discipline, Subjectivity and Personality: An Analysis of the Manuals of Four Psychological Tests'. *History of the Human Sciences* 14, no. 1 (February 2001): 25–47.

'People as Scientific Instruments'. *Spontaneous Generations* 4, no. 1 (2010): 21–29.

Derksen, Maarten, and Eric Rietzschel. 'Surveillance Is Not the Answer, and Replication Is Not a Test: Comment on Kepes and McDaniel, "How Trustworthy Is the Scientific Literature in I-O Psychology?"' *Industrial and Organizational Psychology* 6, no. 3 (2013): 295–98.

Derksen, M., and A. Beaulieu. 'Social Technology'. In *The Sage Handbook of the Philosophy of Social Science*, edited by I. Jarvie and J. Zamora-Bonilla, 703–20. London: Sage, 2011.

Derksen, M., S. Vikkelsø and A. Beaulieu. 'Social Technologies: Cross-Disciplinary Reflections on Technologies in and from the Social Sciences'. *Theory & Psychology* 22, no. 2 (April 1, 2012): 139–47.

Deutsch, M. 'Field Theory in Social Psychology'. In *The Handbook of Social Psychology*, edited by G. Lindzey and E. Aronson, 412–87. Reading, MA: Addison-Wesley, 1968.

Dewey, F. A. 'An Application of Statistical Method'. *American Journal of Sociology* 21, no. 3 (1 November 1, 1915): 334–38.

Dijksterhuis, Ap. 'Welcome Back Theory!' *Perspectives on Psychological Science* 9, no. 1 (January 1, 2014): 72–75.

Dijksterhuis, Ap, and Ad van Knippenberg. 'The Relation between Perception and Behavior, or How to Win a Game of Trivial Pursuit'. *Journal of Personality and Social Psychology* 74, no. 4 (April 1998): 865–77.

Dollard, John. 'Men Who Are Tortured by the Awful Fear of Torture'. *New York Times*. July 29, 1956, sec. *The New York Times Book Review*.

Doyen, Stéphane, Olivier Klein, Cora-Lise Pichon and Axel Cleeremans. 'Behavioral Priming: It's All in the Mind, but Whose Mind?' *PLoS ONE* 7, no. 1 (January 18, 2012): e29081.

Drever, J. 'The Human Factor in Industrial Relations'. In *Industrial Psychology*, edited by Charles Samuel Myers, 16–38. London: T. Butterworth Ltd, 1929.

Duncan, J. 'Efficiency – Real, Unreal, and Brutal'. *American Federationist* 18, no. 5 (1911): 380–84.

Dutton, Kevin. *Flipnosis: The Art of Split-Second Persuasion*. London: William Heinemann, 2010.

Eakin, Hugh, and Mark Danner. 'Our New Politics of Torture'. *NYRBlog*, December 30, 2014. www.nybooks.com/blogs/nyrblog/2014/dec/30/new-politics-torture/.

'Een luisterend oor is het halve werk. Omgaan met opstandigheid'. *BWthema*, no. 5 (2011): 6–7.

Ehrenreich, Barbara. *Bright-Sided: How Positive Thinking Is Undermining America*. New York: Picador, 2010.

Eliot, Thomas D. 'The Social Worker's Criticisms of Undergraduate Sociology'. *Journal of Social Forces* 2, no. 4 (May 1, 1924): 506–12.

Ellwood, Charles. 'Sociology: Its Problems and Its Relations'. *The American Journal of Sociology* 13, no. 3 (November 1907): 300–347.

Elteren, M. van. 'From Emancipating to Domesticating the Workers: Lewinian Social Psychology and the Study of the Work Process till 1947'. In *Recent Trends in Theoretical Psychology*, edited by Henderikus J. Stam, L. P. Mos, W. Thorngate and B. Kaplan, Vol. III. New York: Springer-Verlag, 1993.

Esenwein, J. Berg, and Dale Carnagey. *The Art of Public Speaking*. Springfield: The Home Correspondence School, 1915.

Etzemüller, Thomas, ed. *Die Ordnung der Moderne: Social Engineering im 20. Jahrhundert*. Bielefeld: Transcript, 2009.

'*Social engineering* als Verhaltenslehre des kühlen Kopfes'. In *Die Ordnung der Moderne: Social Engineering im 20. Jahrhundert*, edited by Thomas Etzemüller, 11–39. Bielefeld: Transcript, 2009.

Evans, H.A. 'What Is to Become of the Machinist?' *American Machinist* 33 (December 15, 1910): 1095.

Feenberg, A. 'Looking Backward, Looking Forward. Reflections on the Twentieth Century'. In *Globalization, Technology, and Philosophy*, edited by David Tabachnick and Toivo Koivukoski, 93–105. Albany: State University of New York Press, 2004.

Ferguson, C. J., and M. Heene. 'A Vast Graveyard of Undead Theories: Publication Bias and Psychological Science's Aversion to the Null'. *Perspectives on Psychological Science* 7, no. 6 (2012): 555–61.

Fiedler, Klaus, Florian Kutzner and Joachim I. Krueger. 'The Long Way from α-Error Control to Validity Proper: Problems with a Short-Sighted False-Positive Debate'. *Perspectives on Psychological Science* 7, no. 6 (November 1, 2012): 661–69.

Fitzsimons, G. M., T. L. Chartrand and G. J. Fitzsimons. 'Automatic Effects of Brand Exposure on Motivated Behavior: How Apple Makes You "Think Different"'. *Journal of Consumer Research* 35, no. 1 (June 2008): 21–35.

Follett, Mary. *Dynamic Administration: The Collected Papers of Mary Parker Follett*. Edited by Elliot M. Fox and Lyndall F. Urwick. 2nd ed. New York: Hippocrene Books, 1982.

'The Giving of Orders'. In *Dynamic Administration: The Collected Papers of Mary Parker Follett*, edited by Elliot M. Fox and Lyndall F. Urwick, 2nd ed., 21–41. New York: Hippocrene Books, 1982.

'The Process of Control'. In *Papers on the Science of Administration*, edited by Luther Halsey Gulick and Lyndall F. Urwick, 161–69. New York: Institute of Public Administration, Columbia University, 1937.

Foucault, M. *Discipline and Punish: The Birth of the Prison*. Translated by Alan Sheridan. New ed. London: Penguin, 1991.

'Technologies of the Self'. In *Technologies of the Self: A Seminar with Michel Foucault*, edited by Luther H. Martin, Huck Gutman and Patrick H. Hutton, 16–49. Amherst: University of Massachusetts Press, 1988.

The History of Sexuality. Volume 1: An Introduction. Translated by R. Hurley. Vol. 1. 3 vols. Harmondsworth: Penguin, 1979.

Freedman, Anne, and P. E. Freedman. 'Mass Line Leadership: Another View of Mao's China'. *American Psychologist American Psychologist* 37, no. 1 (1982): 103–5.

French, John R. P., Jr. 'Role-Playing as a Method of Training Foremen'. *Sociometry* 8, no. 3/4 (August 1, 1945): 172–87.

Frey, J.P. 'The Relation of Scientific Management to Labor'. *American Federationist* 20 (1913): 296–302.

Friedmann, Georges. *Industrial Society; the Emergence of the Human Problems of Automation*. Glencoe, IL: Free Press, 1955.

Frison, Guido. 'Some German and Austrian Ideas on "Technologie" and "Technik" between the End of the Eighteenth Century and the Beginning of the Twentieth'. *History of Economic Ideas* 6, no. 1 (January 1, 1998): 107–33.

Gelman, A. 'Primed to Lose'. *Statistical Modeling, Causal Inference, and Social Science*, February 9, 2016. http://andrewgelman.com/2016/02/09/28905/.

Gelman, A., and E. Loken. 'The Statistical Crisis in Science'. *American Scientist* 102, no. 6 (December 2014): 460–65.

Genter, Robert. '"Hypnotizzy" in the Cold War: The American Fascination with Hypnotism in the 1950s'. *The Journal of American Culture* 29, no. 2 (2006): 154–69.

Gergen, K. J. 'Social Psychology as History'. *Journal of Personality and Social Psychology* 26, no. 2 (May 1973): 309–20.

Giddings, Franklin Henry. 'Relation of Sociology to Scientific Studies'. *Journal of Social Science* 32 (November 1, 1894).

'Sociology as a University Study'. *Political Science Quarterly* 6, no. 4 (December 1, 1891): 635–55.

The Principles of Sociology: An Analysis of the Phenomena of Association and of Social Organization. New York: Macmillan Co., 1896.

Gilbreth, Frank. *Primer of Scientific Management ... With an Introduction by Louis D. Brandeis*. London: Constable & Co., 1912.

Gilbreth, L. M. *The Psychology of Management: The Function of the Mind in Determining, Teaching and Installing Methods of Least Waste*. New York: MacMillan Company, 1921.

Gillespie, Richard. *Manufacturing Knowledge: A History of the Hawthorne Experiments*. Cambridge; New York: Cambridge University Press, 1991.

Gompers, S. 'Editorial'. *American Federationist* 18, no. 2 (1911): 116–17.

'Speeding Up'. *American Federationist* 20 (1913): 524.

'The Miracles of Efficiency'. *American Federationist* 18, no. 4 (1911): 273–79.

Gorgias. *Encomium of Helen*. Translated by Douglas M. MacDowell. Bristol [Gloucestershire]: Bristol Classical, 1982.

Gorny, Tom, and Tijn van Ewijk. *MasterFlirt*. Amsterdam: Meulenhoff Boekerij B.V., 2011.

Graebner, William. 'The Small Group and Democratic Social Engineering, 1900–1950'. *Journal of Social Issues* 42, no. 1 (April 1, 1986): 137–54.

Griffin, G. Edward. *Seduction of a Generation – Sensitivity Training as Brainwashing*, 1969. www.youtube.com/watch?v=2gc4g-HCan0.

Guinn, Jeff. *Manson: The Life and Times of Charles Manson*. New York: Simon & Schuster, 2013.

Hacohen, Malachi Haim. *Karl Popper, the Formative Years, 1902–1945: Politics and Philosophy in Interwar Vienna*. Cambridge: Cambridge University Press, 2000.

'Karl Popper, the Vienna Circle, and Red Vienna'. *Journal of the History of Ideas* 59, no. 4 (October 1, 1998): 711–34.

Hansson, Sven Ove. 'Falsificationism Falsified'. *Foundations of Science* 11, no. 3 (September 2006): 275–86.

Harris, Christine R., Noriko Coburn, Doug Rohrer and Harold Pashler. 'Two Failures to Replicate High-Performance-Goal Priming Effects'. *PLoS ONE* 8, no. 8 (August 16, 2013): e72467.

Haskell, Thomas L. *The Emergence of Professional Social Science: The American Social Science Association and the Nineteenth-Century Crisis of Authority*. Urbana: University of Illinois Press, 1977.

Hayes, Edward Cary. 'Sociology as Ethics'. *American Journal of Sociology* 24, no. 3 (November 1, 1918): 289–302.

Hayles, N. Katherine. 'Boundary Disputes: Homeostasis, Reflexivity, and the Foundations of Cybernetics'. *Configurations* 2, no. 3 (1994): 441.

Helmer, Olaf, Bernice B. Brown and Theodore J. Gordon. *Social Technology*. New York: Basic Books, 1966.

Henderson, C. R. 'Applied Sociology (or Social Technology)'. *American Journal of Sociology* 18, no. 2 (1912): 215–21.

'Definition of a Social Policy Relating to the Dependent Group'. *American Journal of Sociology* 10, no. 3 (November 1, 1904): 315–34.

'Review: [Untitled]'. *Journal of Political Economy* 3, no. 2 (March 1, 1895): 236–38.

Social Elements. Institutions, Character, Progress. New York: C. Scribner's Sons, 1898.

'The Scope of Social Technology'. *The American Journal of Sociology* 6, no. 4 (January 1901): 465–86.

Highhouse, S. 'A History of the T-Group and Its Early Applications in Management Development'. *Group Dynamics: Theory, Research, and Practice: The Official Journal of Division 49, Group Psychology and Group Psychotherapy of the American Psychological Association* 6, no. 4 (2002): 277–90.

Hinkle, L. E. 'Communist Manipulation of Behavior. The Russians use Confession to Convict and Condemn, the Chinese to Rehabilitate and Reform'. *Science*, 133 (1961): 1912–14.

Hinkle, L. E., and H. G. Wolff. 'Brainwashing from a Psychological Viewpoint'. Washington, DC: CIA, January 3, 1956.

'Communist Control Techniques: An Analysis of the Methods Used by Communist State Police in the Arrest, Interrogation, and Indoctrination of Persons Regarded as "Enemies of the State"'. Washington, DC: CIA, April 2, 1956.

'Communist Interrogation and Indoctrination of "Enemies of the State"; Analysis of Methods Used by the Communist State Police (a Special Report)'. *A.M.A. Archives of Neurology and Psychiatry* 76, no. 2 (1956): 115–74.

'The Methods of Interrogation and Indoctrination Used by the Communist State Police'. *Bulletin of the New York Academy of Medicine* 33, no. 9 (September 1957): 600–15.

Hogan, Kevin, and James Speakman. *Covert Persuasion: Psychological Tactics and Tricks to Win the Game*. Hoboken, NJ: John Wiley & Sons, 2006.

Hughes, Thomas Parke. *American Genesis: A Century of Invention and Technological Enthusiasm, 1870–1970*. Chicago: University of Chicago Press, 2004.

Hunter, Edward. *Brain-Washing in Red China: The Calculated Destruction of Men's Minds*. New York: Vanguard Press, 1951.

Brainwashing; the Story of Men Who Defied It. New York: Farrar, Straus and Cudahy, 1956.

James, William. *The Principles of Psychology*. Edited by Frederick Burkhardt, Fredson Bowers and Ignas K. Skrupskelis. Cambridge, MA: Harvard University Press, 1981.

Janis, Irving Lester. 'Are the Cominform Countries Using Hypnotic Techniques to Elicit Confessions in Public Trials?' U.S. Airforce Project Rand Research Memorandum. Santa Monica: RAND Corporation, 1949. www.rand.org/pubs/research_memoranda/RM161.html.

John, Leslie, George F. Loewenstein and Drazen Prelec. 'Measuring the Prevalence of Questionable Research Practices with Incentives for Truth-Telling'. *Psychological Science* 23, no. 5 (2012): 524–32.

John, Matthias, George Eckardt and Hans Hiebsch. 'Kurt Lewin's Early Intentions (Dedicated to His 100th Birthday)'. *European Journal of Social Psychology* 19, no. 2 (March 1, 1989): 163–69.

Jordan, John M. *Machine-Age Ideology: Social Engineering and American Liberalism, 1911–1939*. Chapel Hill: University of North Carolina Press, 1994.

Judd, Charles M., and Bertram Gawronski. 'Editorial Comment'. *Journal of Personality and Social Psychology* 100, no. 3 (2011): 406.

Kang, Minsoo. *Sublime Dreams of Living Machines: The Automaton in the European Imagination*. Cambridge, MA: Harvard University Press, 2011.

Kanigel, Robert. *The One Best Way: Frederick Winslow Taylor and the Enigma of Efficiency*. New York: Viking, 1997.

Karlsen, Mads Peter, and Kaspar Villadsen. 'Who Should Do the Talking? The Proliferation of Dialogue as Governmental Technology'. *Culture and Organization* 14, no. 4 (2008): 345–63.

Karremans, Johan C., Wolfgang Stroebe and Jasper Claus. 'Beyond Vicary's Fantasies: The Impact of Subliminal Priming and Brand Choice'. *Journal of Experimental Social Psychology* 42, no. 6 (November 2006): 792–98.

Kaufman, Bruce E. *Managing the Human Factor: The Early Years of Human Resource Management in American Industry*. Ithaca: Cornell University Press, 2008.

Kelly, J. G., J. Dalton and V. T. Francisco. *Interview of Alvin Zander*, 1995. https://vimeo.com/69632837.

Kennedy, George Alexander. *The Art of Persuasion in Greece*. Princeton, NJ: Princeton University Press, 1963.

Kennedy, Raymond, and Ruby Jo Reeves Kennedy. 'Sociology in American Colleges'. *American Sociological Review* 7, no. 5 (October 1, 1942): 661–75.

Killen, Andreas. 'Homo Pavlovius: Cinema, Conditioning, and the Cold War Subject'. *Grey Room*, no. 45 (Fall 2011): 42–59.

Killen, Andreas, and Stefan Andriopoulos. 'Editors' Introduction on Brainwashing: Mind Control, Media, and Warfare'. *Grey Room*, no. 45 (Fall 2011): 7–17.

Klatzky R. L., and J. D. Creswell. 'An Intersensory Interaction Account of Priming Effects – and Their Absence'. *Perspectives on Psychological Science* 9, no. 1 (2014): 49–58.

Klein, Richard A., Kate A. Ratliff, Michelangelo Vianello, Reginald B. Adams Jr., Štěpán Bahník, Michael J. Bernstein, Konrad Bocian, et al. 'Investigating Variation in Replicability: A "Many Labs" Replication Project'. *Social Psychology* 45, no. 3 (2014): 142–52.

Kleinman, S. 'KUBARK Counterintelligence Interrogation Review: Observations of an Interrogator – Lessons Learned and Avenues for Further Research'. In *Educing Information – Interrogation: Science and Art, Foundations for the Future*, 95–140. Washington DC: National Defence Intelligence College, 2006. www.dtic.mil/cgi-bin/GetTRDoc?Location=U2&doc=GetTRDoc.pdf&AD=ADA476636.

Kline, Ronald. 'Construing 'Technology' as "Applied Science": Public Rhetoric of Scientists and Engineers in the United States, 1880–1945'. *Isis* 86, no. 2 (June 1, 1995): 194–221.

Konnikova, Maria. 'Trying to Cure Depression, but Inspiring Torture'. *The New Yorker*, January 14, 2015. www.newyorker.com/science/maria-konnikova/theory-psychology-justified-torture.

Korn, James H. *Illusions of Reality: A History of Deception in Social Psychology*. Albany: State University of New York Press, 1997.

Kuna, David P. 'The Concept of Suggestion in the Early History of Advertising Psychology'. *Journal of the History of the Behavioral Sciences* 12, no. 4 (1976): 347–53.

Kunert, Richard. 'Internal Conceptual Replications Do Not Increase Independent Replication Success'. *Psychonomic Bulletin & Review* 23, no.5 (2016): 1631–38.

Lamme, Victor A. F. *De vrije wil bestaat niet: over wie er echt de baas is in het brein*. Amsterdam: Bakker, 2010.

Lancaster, Jane. *Making Time: Lillian Moller Gilbreth, A Life Beyond 'Cheaper by the Dozen'*. Boston: Northeastern University Press, 2004.

Latour, B. *Reassembling the Social: An Introduction to Actor-Network-Theory*. Clarendon Lectures in Management Studies. Oxford: Oxford University Press, 2005.

The Pasteurization of France. Translated by A. Sheridan and Law. Cambridge, MA: Harvard University Press, 1988.

'*The Prince* for Machines as Well as for Machinations'. In *Technology and Social Process*, edited by B. Elliott, 20–43. Edinburgh: Edinburgh University Press, 1988.

'Visualization and Cognition: Thinking with Eyes and Hands'. In *Knowledge and Society: Studies in the Sociology of Culture Past and Present*, edited by H. Kuklick, 6:1–40. JAI Press, 1986.

'When Things Strike Back: A Possible Contribution of "Science Studies" to the Social Sciences'. *British Journal of Sociology* 51, no. 1 (2000): 107–23.

'Where Are the Missing Masses? The Sociology of a Few Mundane Artifacts'. In *Shaping Technology/Building Society. Studies in Sociotechnical Change*, 225–58. Cambridge, MA: MIT Press, 1992.

Lehrer, Jonah. 'The Truth Wears Off'. *The New Yorker*, December 13, 2010. www.newyorker.com/magazine/2010/12/13/the-truth-wears-off.

Lemov, Rebecca. 'Brainwashing's Avatar: The Curious Career of Dr. Ewen Cameron'. *Grey Room*, no. 45 (Fall 2011): 61–87.

Leopold, Jason. 'Psychologist James Mitchell Admits He Waterboarded Al Qaeda Suspects'. *VICE News*, December 15, 2014. https://news.vice.com/article/psychologist-james-mitchell-admits-he-waterboarded-al-qaeda-suspects.

Levelt, W. J. M., M. S. Groenhuijsen and J. A. P. Hagenaars. 'Interim-rapportage inzake door prof. dr. D.A. Stapel gemaakte inbreuk op wetenschappelijke integriteit'. Tilburg: Universiteit van Tilburg, October 31, 2011.

Lewin, Kurt. *Die Sozialisierung des Taylorsystems: eine grundsätzliche Untersuchung zur Arbeits- u. Berufs-Psychologie*. Vol. 4. Praktischer Sozialismus. Berlin: Gesellschaft und Erziehung, 1920.

'Frontiers in Group Dynamics. Concept, Method and Reality in Social Science; Social Equilibria and Social Change'. *Human Relations* 1 (1947): 5–41.

'The Relative Effectiveness of a Lecture Method and a Method of Group Decision for Changing Food Habits'. Washington, DC: Committee on Food Habits, National Research Council, 1942. Box 948, folder #56. The Drs. Nicholas and Dorothy Cummings Center for the History of Psychology.

'The Research Center for Group Dynamics at Massachusetts Institute of Technology'. *Sociometry* 8, no. 2 (1945): 126–36.

'The Solution of a Chronic Conflict in a Factory'. In *Proceedings of the Second Brief Psychotherapy Council*, 36–46. Chicago, 1944.

'The Solution of a Chronic Conflict in a Factory. Analysis of a Report by Alex Bavelas'. Iowa, undated. Box M948, folder #57. The Drs. Nicholas and Dorothy Cummings Center for the History of Psychology.

'The Solution of a Chronic Conflict in Industry (1944)'. *In Resolving Social Conflicts and Field Theory in Social Science.*, 94–104. Washington, DC: American Psychological Association, 1997.

Lewis, Sinclair. 'Car-Yes-Man'. *Newsweek*, November 15, 1937.

'One Man Revolution'. *Newsweek*, November 22, 1937.

Lezaun, J. 'A Market of Opinions: The Political Epistemology of Focus Groups'. *Sociological Review* 55 (October 2007): 130–51.

Lezaun, Javier, and Nerea Calvillo. 'In the Political Laboratory: Kurt Lewin's Atmospheres'. *Journal of Cultural Economy* 7, no. 4 (2014): 434–57.

'Lezingen en workshops'. *PACELLEVANGOETHEM*. Accessed September 4, 2014. http://pacellevangoethem.nl/lezing-workshop/.

Lieberman, Matthew D. 'Latitudes of Acceptance. A Conversation with Matthew D. Lieberman'. *Edge.org*, 8 2014. http://edge.org/conversation/latitudes-of-acceptance.

Lifton, Robert J. 'Chinese Communist "Thought Reform": Confession and Re-Education of Western Civilians'. *Bulletin of the New York Academy of Medicine* 33, no. 9 (September 1957): 626–44.

'"Thought Reform" of Western Civilians in Chinese Communist Prisons'. *Psychiatry* 19, no. 2 (May 1, 1956): 173–95.

Thought Reform and the Psychology of Totalism: A Study of 'Brainwashing' in China. Chapel Hill: University of North Carolina Press, 1961.

Lind, Andrew W. 'Some Ecological Patterns of Community Disorganization in Honolulu'. *American Journal of Sociology* 36, no. 2 (September 1, 1930): 206–20.

Lippitt, Ronald. *Training in Community Relations; a Research Exploration toward New Group Skills*. New York: Harper, 1949.

Locke, E. A. 'Theory Building, Replication, and Behavioral Priming: Where Do We Need to Go from Here?' *Perspectives on Psychological Science* 10, no. 3 (May 18, 2015): 408–14.

Lopes, Lola L. 'The Rhetoric of Irrationality'. *Theory & Psychology* 1, no. 1 (February 1, 1991): 65–82.

Lubove, Roy. *The Professional Altruist: The Emergence of Social Work as a Career, 1880–1930*. A Publication of the Center for the Study of the History of Liberty in America, Harvard University. Cambridge, MA: Harvard University Press, 1968.

Lutz, Catherine. 'Epistemology of the Bunker: The Brainwashed and Other New Subjects of Permanent War'. In *Inventing the Psychological: Toward a Cultural History of Emotional Life in America*, edited by Joel Pfister and Nancy Schnog, 245–67. New Haven: Yale University Press, 1997.

Machiavelli, Niccolò. *The Prince*. Translated by Russell Price. Cambridge; New York: Cambridge University Press, 1988.

Maier, Charles S. 'Between Taylorism and Technocracy: European Ideologies and the Vision of Industrial Productivity in the 1920s'. *Journal of Contemporary History* 5, no. 2 (January 1, 1970): 27–61.

Marklund, Carl. 'Begriffsgeschichte and Übergriffsgeschichte in the History of Social Engineering'. In *Die Ordnung der Moderne: Social Engineering im 20. Jahrhundert*, edited by Thomas Etzemüller. Bielefeld: Transcript, 2009.

Marks, John D. *The Search for the 'Manchurian Candidate': The CIA and Mind Control*. London: A. Lane, 1979.

Marrow, Alfred Jay. *The Practical Theorist: The Life and Work of Kurt Lewin*. New York: Basic Books, 1969.

Marx, Karl. *Capital: A Critique of Political Economy – Vol. I-Part I: The Process of Capitalist Production*. New York: Cosimo, Inc., 2007.

Marx, Leo. 'Technology: The Emergence of a Hazardous Concept'. *Technology and Culture* 51, no. 3 (2010): 561–77.

Mayo, Charles W., and United States Delegation to the General Assembly. *Statement by The Honorable Charles W. Mayo, United States Alternate Representative on the Political Committee on 'The Question of Impartial Investigation of Charges of Use by United Nations Forces of Bacteriological Warfare'*. Washington: US Delegation to the General Assembly, 1953.

McCauley, Clark. 'More about Thought Reform'. *American Psychologist* 37, no. 1 (January 1982): 102–3.

McComiskey, Bruce. *Gorgias and the New Sophistic Rhetoric*. Carbondale: Southern Illinois University Press, 2002.

McCoy, Alfred. *A Question of Torture. CIA Interrogation, from the Cold War to the War on Terror*. New York: Henry Holt and Company, 2006.

McGee, Micki. *Self-Help, Inc.: Makeover Culture in American Life*. Oxford; New York: Oxford University Press, 2005.

McGregor, Douglas Murray. 'The Human Side of Enterprise'. *Management Review* 46, no. 11 (November 1, 1957).

Meerloo, J. A. M. *Een mond vol spijkers: een psycholoog op oorlogspad*. Wassenaar: Servire, 1975.

'Pavlovian Strategy as a Weapon of Menticide'. *The American Journal of Psychiatry* 110, no. 11 (1954): 809–13.

'The Crime of Menticide'. *American Journal of Psychiatry* 107, no. 8 (February 1, 1951): 594–98.

The Rape of the Mind: The Psychology of Thought Control, Menticide, and Brainwashing. Cleveland, OH: The World Publishing Company, 1956.

Melley, Timothy. 'Brain Warfare: The Covert Sphere, Terrorism, and the Legacy of the Cold War'. *Grey Room*, no. 45 (Fall 2011): 19–40.

'Brainwashed! Conspiracy Theory and Ideology in the Postwar United States'. *New German Critique* 35, no. 103 (March 20, 2008): 145–64.

Merton, Robert K. *Social Theory and Social Structure*. Enlarged ed. New York: Free Press, 1968.

Meyer, Donald. *The Positive Thinkers. A Study of the American Quest for Health, Wealth and Personal Power from Mary Baker Eddy to Norman Vincent Peale*. New York: Doubleday, 1965.

Miller, Carolyn. 'Should We Name the Tools?' In *The Public Work of Rhetoric: Citizen-Scholars and Civic Engagement*, edited by John Ackerman and David Coogan, 19–38. Studies in Rhetoric/Communication. Columbia: University of South Carolina Press, 2010.

Miller, Irving. 'Letter from Rabbi Irving Miller, Chairman of the Executive Committee of the American Jewish Congress, to Karl Compton, President of MIT', November 16, 1944. Box M293, folder #11. The Drs. Nicholas and Dorothy Cummings Center for the History of Psychology.

Miller, James G. 'Brainwashing: Present and Future'. *Journal of Social Issues* 13, no. 3 (July 1, 1957): 48–55.

Mitchell, Jason. 'On the Emptiness of Failed Replications'. Accessed August 28, 2014. https://web.archive.org/web/20150604192510/http://wjh.harvard.edu/~jmitchel/writing/failed_science.htm.

Mönning, Paul Hassels. 'Neuromarketing: De "koopknop" is een illusie'. *Marketing Online*, October 30, 2012. www.marketingonline.nl/bericht/neuromarketing-de-koopknop-is-een-illusie.

Morawski, Jill. 'Epistemological Dizziness in the Psychology Laboratory: Lively Subjects, Anxious Experimenters, and Experimental Relations, 1950–1970'. *Isis* 106, no. 3 (September 1, 2015): 567–97.

Moreno, Jonathan D. *Mind Wars: Brain Science and the Military in the Twenty-First Century*. New York: Bellevue Literary Press, 2012.

Münsterberg, H. *On the Witness Stand: Essays in Psychology and Crime*. New York: Doubleday, Page & Company, Inc., 1908.

Psychology and Industrial Efficiency. Boston: Houghton Mifflin, 1913.

'The Market and Psychology'. In *American Problems.*, 151–73. New York: Moffat, Yard & Co., 1910.

Myers, C. S. *Mind and Work: The Psychological Factors in Industry and Commerce*. London: University of London Press, 1927.

Present-Day Applications of Psychology, with Special Reference to Industry, Education and Nervous Breakdown. London: Methuen, 1918.

'The Efficiency Engineer and the Industrial Psychologist'. *Journal of the National Institute of Industrial Psychology* 1, no. 5 (1923): 168–72.

Nadworny, Milton J. *Scientific Management and the Unions, 1900–1932: A Historical Analysis*. Cambridge, MA: Harvard University Press, 1955.

Natorp, Paul. *Gesammelte Abhandlungen zur Sozialpädagogik*. Stuttgart: Frommans, 1922.

Sozialpädagogik: Theorie der Willenserziehung auf der Grundlage der Gemeinschaft. Stuttgart: Fr. Frommanns Verlag (E. Hauff), 1899.

Nelson, Daniel. *Frederick W. Taylor and the Rise of Scientific Management*. Madison: University of Wisconsin Press, 1980.

'Scientific Management, Systematic Management, and Labor, 1880–1915'. *The Business History Review* 48, no. 4 (December 1, 1974): 479–500.

Neurensics and dijksterhuis&vanbaaren. 'Front Runners in Neuro-Economie Kondigen Alliantie Aan'. Press release. Amsterdam, July 6, 2012. www.db-abs.com/downloads/120706_persbericht%20alliantie.pdf.

Neuroskeptic. 'The Replication Crisis: Response to Lieberman'. *Neuroskeptic*, August 31, 2014. http://blogs.discovermagazine.com/neuroskeptic/2014/08/31/replication-crisis-response-lieberman/.

'NeuroStandards Project White Paper'. Advertising Research Foundation, October 2011.

Nicholson, Ian. '"Torture at Yale": Experimental Subjects, Laboratory Torment and The "rehabilitation" of Milgram's "Obedience to Authority"'. *Theory & Psychology* 21, no. 6 (2011): 737–61.

Nosek, B. A., and Y. Bar-Anan. 'Scientific Utopia: I. Opening Scientific Communication'. *Psychological Inquiry* 23, no. 3 (July 2012): 217–43.

Nosek, B. A., J. R. Spies and M. Motyl. 'Scientific Utopia: II. Restructuring Incentives and Practices to Promote Truth Over Publishability'. *Perspectives on Psychological Science* 7, no. 6 (2012): 615–31.

Odum, Howard W. *American Sociology: The Story of Sociology in the United States through 1950*. New York: Greenwood Press, 1951.

Ofner, Julius. *Studien sozialer Jurisprudenz*. Vienna: Hölder, 1894.

Ogburn, William Fielding. 'Technology and Sociology'. *Social Forces* 17, no. 1 (October 1, 1938): 1–8.

'The Folkways of a Scientific Sociology'. *Publications of the American Sociological Society* 24 (1930): 1–11.

Oldenziel, Ruth. *Making Technology Masculine: Men, Women and Modern Machines in America, 1870–1945*. Amsterdam: Amsterdam University Press, 1999.

Open Science Collaboration. 'An Open, Large-Scale, Collaborative Effort to Estimate the Reproducibility of Psychological Science'. *Perspectives on Psychological Science* 7, no. 6 (November 1, 2012): 657–60.

Overstreet, H. A. *Influencing Human Behavior*. New York: W.W. Norton, 1925.

Owen, Robert. *A New View of Society. Or: Essays on the Formation of the Human Character Preparatory to the Development of a Plan for Gradually Ameliorating the Condition of Mankind*. 2nd ed. Clifton, NJ: A.M. Kelley, 1972.

Oxtoby, Roby. 'An Observational Method for Determining the Job of Individual Floor Girls', undated. Box M1940, folder #8. The Drs. Nicholas and Dorothy Cummings Center for the History of Psychology.

Pashler, Harold, Noriko Coburn and Christine R. Harris. 'Priming of Social Distance? Failure to Replicate Effects on Social and Food Judgments'. *PLoS ONE* 7, no. 8 (August 29, 2012): e42510.

'Pastor Raps Best Seller. Asserts Dale Carnegie's Book Is a Mockery of True Friendship'. *Reading Eagle*, March 18, 1938.

Patnoe, Shelley. *A Narrative History of Experimental Social Psychology: The Lewin Tradition*. New York: Springer-Verlag, 1988.

Paul, Joanne. 'The Use of Kairos in Renaissance Political Philosophy'. *Renaissance Quarterly* 67, no. 1 (March 1, 2014): 43–78.

Peck, E.C. 'Systematic versus Scientific Management. A Word of Caution against Precipitate Introduction of Radical Shop-Working Measures'. *Iron Age* 88 (August 17, 1911): 364–65.

Pels, Dick, Kevin Hetherington and Frédéric Vandenberghe. 'The Status of the Object Performances, Mediations, and Techniques'. *Theory, Culture & Society* 19, no. 5–6 (December 1, 2002): 1–21.

Pettit, Michael. *The Science of Deception: Psychology and Commerce in America*. Chicago; London: University of Chicago Press, 2013.

Pickel, A. 'Between Social Science and Social Technology – Toward a Philosophical Foundation for Post-Communist Transformation Studies'. *Philosophy of the Social Sciences* 31, no. 4 (December 2001): 459–87.

Pinch, T., M. Ashmore and M. Mulkay. 'Technology, Testing Text: Clinical Budgeting in the U.K. National Health Service'. In *Shaping Technology/Building Society. Studies in Sociotechnical Change*, edited by W. E. Bijker and J. Law, 265–89. Cambridge, MA: MIT Press, 1992.

Ping, Hu. *The Thought Remolding Campaign of the Chinese Communist Party-State*. Amsterdam: Amsterdam University Press, 2012.

Plato. *The Collected Dialogues of Plato: Including the Letters*. Edited by Edith Hamilton and Huntington Cairns. Princeton, NJ: Princeton University Press, 1963.

Popper, Karl. 'The Poverty of Historicism, I'. *Economica*, New Series, 11, no. 42 (May 1, 1944): 86–103.

'The Poverty of Historicism, II. A Criticism of Historicist Methods'. *Economica*, New Series, 11, no. 43 (1944): 119–37.

'The Poverty of Historicism, III'. *Economica*, New Series, 12, no. 46 (May 1, 1945): 69–89.

The Open Society and Its Enemies. Volume I: The Spell of Plato. 5th ed. Vol. 1. 2 vols. London: Routledge & Kegan Paul, 1966.

The Open Society and Its Enemies. Volume II. The High Tide of Prophecy: Hegel, Marx, and the Aftermath. 5th ed. London: Routledge & Kegan Paul, 1966.

Unended Quest: An Intellectual Autobiography. London: Routledge, 1992.

Post, Julius. *Arbeit statt Almosen: Beitrag zur Social-Technik*. Bremen: Roussell, 1881.

Pratkanis, A. R. 'The Cargo-Cult Science of Subliminal Persuasion – CSI'. *Skeptical Enquirer* 16, no. 3 (1992). www.csicop.org/si/show/cargo-cult_science_of_subliminal_persuasion.

'Preliminary Report of the First National Training Laboratory on Group Development: Sponsored by the National Education Association and

Research Center for Group Dynamics, M.I.T. Bethel, Maine, June 16 to July 4, 1947'. National Education Association, 1947.

Price, C. 'Interview with Mason Haire', undated. Box M946, folder #21. The Drs. Nicholas and Dorothy Cummings Center for the History of Psychology.

'Private Institutions Used in C.I.A. Effort to Control Behavior; 25-Year, $25 Million Program New Information About Funding and Operations Disclosed by Documents and Interviews Private Institutions Used in C.I.A. Plan'. *The New York Times*, August 2, 1977.

Publication of the American Sociological Society. Vol. 8, 1913.

'Pulitzer Critic Rates Top Ten Culture Books'. *Schenectady Gazette*, March 27, 1985.

Quintilian, Marcus Fabius. *The Institutio Oratoria of Quintilian*, translated by Harold Edgeworth Butler. London: Heinemann, 1920–1922.

Radder, H. 'Causality, Theory Ladenness and Reproducibility in Experimental Science' (n.d.). www.academia.edu/26560040/Causality_Theory_Ladenness_and_Reproducibility_in_Experimental_Science.

'De Cognitieve drie-eenheid: realiseren, interpreteren en abstraheren'. *Algemeen Nederlands Tijdschrift Voor Wijsbegeerte* 106, no. 1 (2014): 3–36.

'Science, Technology and the Science–Technology Relationship'. In *Philosophy of Technology and Engineering Sciences*, edited by Anthonie Meijers. Amsterdam; London; Boston: Elsevier/North Holland, 2009.

Ramage, John D. *Twentieth-Century American Success Rhetoric: How to Construct a Suitable Self*. Carbondale: Southern Illinois University Press, 2005.

Ray, L. J. 'Critical Theory and Positivism: Popper and the Frankfurt School'. *Philosophy of the Social Sciences. An International Journal Aberdeen* 9, no. 2 (1979): 149–73.

Riley, Thomas J. 'Sociology and Social Surveys'. *American Journal of Sociology* 16, no. 6 (May 1, 1911): 818–36.

Ritchie, S. J., R. Wiseman and C. C. French. 'Failing the Future: Three Unsuccessful Attempts to Replicate Bem's "Retroactive Facilitation of Recall" Effect'. *PLoS ONE* 7, no. 3 (March 14, 2012): e33423.

Ritchie, S. J., R. Wiseman and C. C. French. 'Replication, Replication, Replication'. *The Psychologist* 25 (May 2012): 346–57.

Rivers, Nathaniel A., and M. Derksen. 'Ecologies of Deception in Psychology and Rhetoric'. *Quarterly Journal of Speech* 101, no. 4 (2016): 633–54.

Rose, N. 'Engineering the Human Soul: Analyzing Psychological Expertise'. *Science in Context* 5, no. 2 (1992): 351–69.

Inventing Our Selves: Psychology, Power, and Personhood. Cambridge Studies in the History of Psychology. Cambridge: Cambridge University Press, 1996.

Ross, Dorothy. *The Origins of American Social Science*. Cambridge; New York: Cambridge University Press, 1991.

Rutherford, Alexandra. *Beyond the Box: B.F. Skinner's Technology of Behavior from Laboratory to Life, 1950s–1970s*. Toronto: University of Toronto Press, 2009.

Sanderson, Dwight. 'Discussion of Sociometry'. *Sociometry* 6, no. 3 (1943): 214–18.

Saroyan, William. *The Trouble with Tigers*. New York: Harcourt, Brace and Co, 1938.

Schachter, H. L. 'Frederick Winslow Taylor and the Idea of Worker Participation. A Brief against Easy Administrative Dichotomies'. *Administration and Society* 21, no. 1 (1989): 20–31.

Schatzberg, Eric. 'Technik Comes to America: Changing Meanings of Technology before 1930'. *Technology and Culture* 47, no. 3 (2006): 486–512.

Scheflin, Alan W., and Edward M. Opton. *The Mind Manipulators: A Non-Fiction Account*. New York: Paddington Press (distributed by Grosset & Dunlap), 1978.

Schein, Edgar H. 'Knuckling Under'. *Contemporary Psychology: A Journal of Reviews* 2, no. 5 (1957): 140–41.

Inge Schneier and Curtis H. Barker*Coercive Persuasion: A Socio-Psychological Analysis of the 'Brainwashing' of American Civilian Prisoners by the Chinese Communists*. New York: Norton, 1961.

'The Chinese Indoctrination Program for Prisoners of War; a Study of Attempted Brainwashing'. *Psychiatry* 19, no. 2 (1956): 149–72.

Schlenker, Barry R. 'Social Psychology and Science'. *Journal of Personality and Social Psychology* 29, no. 1 (1974): 1–15.

Schnall, Simone. 'Moral Intuitions, Replication, and the Scientific Study of Human Nature', 2014. http://edge.org/conversation/simone-schnall-moral-intuitions-replication-and-the-scientific-study-of-human-nature.

'Simone Schnall on Her Experience with a Registered Replication Project'. *Character and Context*, May 23, 2014. www.spspblog.org/simone-schnall-on-her-experience-with-a-registered-replication-project/.

'Social Media and the Crowd-Sourcing of Social Psychology'. Page. *Cambridge Embodied Cognition and Emotion Laboratory Blog*, November 18, 2014. www.psychol.cam.ac.uk/cece/blog.

Schneider, Tanja, and Steve Woolgar. 'Technologies of Ironic Revelation: Enacting Consumers in Neuromarkets'. *Consumption Markets & Culture* 15, no. 2 (2012): 169–89.

Schröder, Tobias, and Paul Thagard. 'The Affective Meanings of Automatic Social Behaviors: Three Mechanisms That Explain Priming'. *Psychological Review* 120, no. 1 (2013): 255–80.

Schwartz, B. 'On Klatzky and Creswell (2014): Saving Social Priming Effects but Losing Science as We Know It?' *Perspectives on Psychological Science* 10, no. 3 (May 18, 2015): 404–7.

Schwarzkopf, Stefan. 'Measurement Devices and the Psychophysiology of Consumer Behaviour: A Posthuman Genealogy of Neuromarketing'. *BioSocieties* 10, no. 4 (December 2015): 465–82.

Scott, Walter Dill. *Increasing Human Efficiency in Business: A Contribution to the Psychology of Business*. New York: Macmillan, 1911.

Influencing Men in Business; the Psychology of Argument and Suggestion. New York: The Ronald Press Company, 1911.

The Psychology of Public Speaking, Adapted for Use as a Text Book in High Schools and Colleges. New York: Noble and Noble, 1926.

Seed, David. *Brainwashing. The Fictions of Mind Control: A Study of Novels and Films since World War II*. Kent, OH: Kent State University Press, 2004.

Senate Select Committee on Intelligence. 'The Senate Intelligence Committee Report on Torture: Committee Study of the Central Intelligence Agency's Detention and Interrogation Program', 2014.

Shanks, David R., Ben R. Newell, Eun Hee Lee, Divya Balakrishnan, Lisa Ekelund, Zarus Cenac, Fragkiski Kavvadia and Christopher Moore. 'Priming Intelligent Behavior: An Elusive Phenomenon'. *PLoS ONE* 8, no. 4 (April 24, 2013): e56515.

Shaw, Tamsin. 'The Psychologists Take Power'. *The New York Review of Books*, February 25, 2016. www.nybooks.com/articles/2016/02/25/the-psychologists-take-power/.

Simkin, Colin George Frederick. *Popper's Views on Natural and Social Science*. BRILL, 1993.

Simmons, Joseph P., Leif D. Nelson and Uri Simonsohn. 'False-Positive Psychology'. *Psychological Science* 22, no. 11 (November 1, 2011): 1359–66.

Simonsohn, U. 'Just Post It: The Lesson from Two Cases of Fabricated Data Detected by Statistics Alone'. *Psychological Science* 24, no. 10 (2013): 1875–88.

Sipiora, Phillip, and James S. Baumlin, eds. *Rhetoric and Kairos Essays in History, Theory, and Praxis*. Albany, NY: State University of New York Press, 2002.

Sloane, Thomas O., ed. *Encyclopedia of Rhetoric*. Oxford: Oxford University Press, 2001.

Small, Albion W. 'Fifty Years of Sociology in the United States (1865–1915)'. *American Journal of Sociology* 21, no. 6 (May 1, 1916): 721–864.

General Sociology. An Exposition of the Main Development in Sociological Theory from Spencer to Ratzenhofer. Chicago: University of Chicago Press, 1905.

'General Sociology'. *American Journal of Sociology* 18, no. 2 (1912): 200–14.

'Points of Agreement among Sociologists'. *American Journal of Sociology* 12, no. 5 (March 1, 1907): 633–55.

'Seminar Notes: The Methodology of the Social Problem. Division I. The Sources and Uses of Material'. *The American Journal of Sociology* 4, no. 1 (July 1898): 113–44.

'The Scope of Sociology. VI. Some Incidents of Association'. *American Journal of Sociology* 6, no. 3 (November 1, 1900): 324–80.

'The Scope of Sociology. VII. Classification of Associations'. *American Journal of Sociology* 6, no. 4 (January 1, 1901): 487–531.

'The Scope of Sociology. IX. Premises of Practical Sociology'. *American Journal of Sociology* 10, no. 1 (July 1, 1904): 26–46.

Small, Albion W., and George Vincent. *An Introduction to the Study of Society*. New York: American Book Co., 1894.

Smith, Mark C. *Social Science in the Crucible: The American Debate over Objectivity and Purpose, 1918–1941*. Durham, NC: Duke University Press, 1994.

'Soft Answers'. *New York Times*. February 14, 1937, sec. Book Review.

'Specific Gripes', undated. Box 220, folder #1. The Drs. Nicholas and Dorothy Cummings Center for the History of Psychology.

Sproule, J. Michael. 'Inventing Public Speaking: Rhetoric and the Speech Book, 1730–1930'. *Rhetoric & Public Affairs* 15, no. 4 (2012): 563–608.

Srivastava, Sanjay. 'An Open Review of Many Labs 3: Much to Learn'. *The Hardest Science*, March 12, 2015. https://hardsci.wordpress.com/2015/03/12/an-open-review-of-many-labs-3-much-to-learn/.

'Making Progress in the Hardest Science'. *The Hardest Science*, March 14, 2009. http://hardsci.wordpress.com/2009/03/14/making-progress-in-the-hardest-science/.

Steiner, Jesse F. *Education for Social Work*. Chicago: University of Chicago Press, 1921.

'The Limitations of a Conceptual Approach to the Applications of Sociology to Social Work'. *Social Forces* 7, no. 4 (June 1, 1929): 500–502.

Story, Christopher. *The New Underworld Order: Triumph of Criminalism the Global Hegemony of Masonic Intelligence*. Westminster: Edward Harle, 2003.

Stroebe, Wolfgang. 'The Subtle Power of Hidden Messages'. *Scientific American Mind* 23, no. 2 (June 5, 2012): 46–51.

Stroebe, Wolfgang, Tom Postmes and Russell Spears. 'Scientific Misconduct and the Myth of Self-Correction in Science'. *Perspectives on Psychological Science* 7, no. 6 (November 1, 2012): 670–88.

Stroebe, Wolfgang, and Fritz Strack. 'The Alleged Crisis and the Illusion of Exact Replication'. *Perspectives on Psychological Science* 9, no. 1 (January 1, 2014): 59–71.

Susman, Warren. *Culture as History: The Transformation of American Society in the Twentieth Century*. New York: Pantheon Books, 1984.

Taylor, Carl C. 'The Social Survey and the Science of Sociology'. *American Journal of Sociology* 25, no. 6 (May 1, 1920): 731–56.

Taylor, Frederick Winslow. 'A Piece-Rate System: Being a Step toward Partial Solution of the Labor Problem'. In *Scientific Management. A Collection of the More Significant Articles Describing the Taylor System of Management*, edited by C. B. Thompson, 636–83. Easton: Hive Pub. Co., 1972.

Scientific Management: Comprising Shop Management, The Principles of Scientific Management and Testimony before the Special House Committee. New York: Harper, 1947.

'Shop Management'. In *Scientific Management: Comprising Shop Management, The Principles of Scientific Management and Testimony before the Special House Committee*. New York: Harper, 1947.

'Testimony before the Special House Committee'. In *Scientific Management: Comprising Shop Management, The Principles of Scientific Management and Testimony before the Special House Committee*. New York: Harper, 1947.

'The Principles of Scientific Management'. In *Scientific Management: Comprising Shop Management, The Principles of Scientific Management and Testimony before the Special House Committee*. New York: Harper, 1947.

Throop, Robert, and Lloyd Gordon Ward. 'Organization of Academic Foci within the "Department of Social Science"', 2006. www.brocku.ca/MeadProject/Timeline/Department_foci_Soci.html.

Thurber, James. 'The Voice with the Smile. Review of Dale Carnegie, How to Win Friends and Influence People'. *Saturday Review*, January 30, 1937.

Tolman, Frank. 'The Study of Sociology in Institutions of Learning in the United States. II'. *The American Journal of Sociology* 8, no. 1 (July 1902): 85–121.

Tressler, Irving D. *How to Lose Friends and Alienate People*. New York: Stackpole Sons, 1937.

US Congress. House. Committee on Labor. *Investigation of Taylor System of Shop Management. Hearings before the Committee on Labor of the House of Representatives, Sixty-Second Congress, First Session, on House Resolution 90 ...* Washington: Government Printing Office, 1911.

House. Committee on Labor. *The Taylor and Other Systems of Shop Management. Hearings before Special Committee of the House of Representatives to Investigate the Taylor and Other Systems of Shop Management under Authority of H. Res. 90 ... [Oct. 4, 1911–Feb. 12, 1912]*. 3 vols. Washington: Government Printing Office, 1912.

Van Asperen de Boer, Rosanne. 'Herziene editie iJs verkopen aan Eskimo's'. *PACELLEVANGOETHEM*, March 10, 2014. http://pacellevangoethem.nl/herziene-editie-ijs-verkopen-aan-eskimos-2/.

Van De Water, Thomas J. 'Psychology's Entrepreneurs and the Marketing of Industrial Psychology'. *Journal of Applied Psychology* 82, no. 4 (1997): 486–99.

Van Ginneken, Jac. *Zielkunde en Taylor-systeem*. Zielkundige verwikkelingen: wegwijzertjes voor eenling en gemeenschap. 1. Amsterdam: De R.K. Boekcentrale, 1918.

Van Goethem, Pacelle. *IJs verkopen aan eskimo's*. Amsterdam: Atlas Contact, Uitgeverij, 2012.

Veblen, Thorstein B. 'Some Neglected Points in the Theory of Socialism'. *Annals of the American Academy of Political and Social Science* 2 (November 1, 1891): 57–74.

The Engineers and the Price System. New York: B. W. Huebsch, Inc., 1921.

Vickers, Brian. *In Defence of Rhetoric*. Oxford: Clarendon Press, 1988.

Vincent, George E. 'Varieties of Sociology'. *American Journal of Sociology* 12, no. 1 (July 1, 1906): 1–10.

Wagenmakers, Eric–Jan, Ruud Wetzels, Denny Borsboom and Han L. J. van der Maas. 'Why Psychologists Must Change the Way They Analyze Their Data: The Case of Psi: Comment on Bem (2011)'. *Journal of Personality and Social Psychology* 100, no. 3 (March 2011): 426–32.

Wallace, Robert W. 'Starting a Department and Getting It Under Way: Sociology at Columbia University, 1891–1914'. *Minerva* 30, no. 4 (December 1, 1992): 497–512.

Watts, Steven. *Self-Help Messiah: Dale Carnegie and Success in Modern America*. New York: Other Press, 2013.

Weil, Simone. 'Prospects. Are We Heading for a Proletarian Revolution?' In *Oppression and Liberty*, 1–23. London: Routledge and Paul, 1958.

Wilde, Rein de. *Discipline en legende: de identiteit van de sociologie in Duitsland en de Verenigde Staten, 1870–1930*. Amsterdam: Van Gennep, 1992.

Williams, R. *Keywords. A Vocabulary of Culture and Society*. London: Fontana, 1988.

Wilson, Andrew. 'Psychology's Real Replication Problem: Our Methods Sections'. *Notes from Two Scientific Psychologists*, May 26, 2014. http://psychsciencenotes.blogspot.co.uk/2014/05/psychologys-real-replication-problem.html.

Yong. E. 'A Failed Replication Draws a Scathing Personal Attack from a Psychology Professor'. *Not Exactly Rocket Science*, March 10, 2012. http://blogs.discovermagazine.com/notrocketscience/2012/03/10/failed-replication-bargh-psychology-study-doyen/.

'Replication Studies: Bad Copy'. *Nature* 485 (May 17, 2012): 298–300.

Young, Charles S. *Name, Rank, and Serial Number: Exploiting Korean War POWs at Home and Abroad*, 2014. Oxford: Oxford University Press.

Index